Advanced Multi-GNSS Positioning and Its Applications in Geoscience

Advanced Multi-GNSS Positioning and Its Applications in Geoscience

Guest Editors

Ahao Wang
Yize Zhang
Xuexi Liu
Xiangdong An
Junping Chen

MDPI

Basel • Beijing • Wuhan • Barcelona • Belgrade • Novi Sad • Cluj • Manchester

Guest Editors

Ahao Wang
College of Geoscience and Surveying Engineering
China University of Mining and Technology-Beijing
Beijing
China

Yize Zhang
Shanghai Astronomical Observatory
Chinese Academy of Sciences
Shanghai
China

Xuexi Liu
School of Environment Science and Spatial Informatics
China University of Mining and Technology
Xuzhou
China

Xiangdong An
School of Electrical and Electronic Engineering
Nanyang Technological University
Singapore
Singapore

Junping Chen
Shanghai Astronomical Observatory
Chinese Academy of Sciences
Shanghai
China

Editorial Office
MDPI AG
Grosspeteranlage 5
4052 Basel, Switzerland

This is a reprint of the Special Issue, published open access by the journal *Remote Sensing* (ISSN 2072-4292), freely accessible at: https://www.mdpi.com/journal/remotesensing/special_issues/F9XFGGOLM3.

For citation purposes, cite each article independently as indicated on the article page online and as indicated below:

Lastname, A.A.; Lastname, B.B. Article Title. *Journal Name* **Year**, *Volume Number*, Page Range.

ISBN 978-3-7258-6636-6 (Hbk)
ISBN 978-3-7258-6637-3 (PDF)
https://doi.org/10.3390/books978-3-7258-6637-3

Contents

remote sensing

MDPI

Article

Exploring the Advantages of Multi-GNSS Ionosphere-Weighted Single-Frequency Precise Point Positioning in Regional Ionospheric VTEC Modeling

Ahao Wang [1], Yize Zhang [2,3,4,*], Junping Chen [2,3,4], Hu Wang [5], Xuexi Liu [6], Yihang Xu [1], Jing Li [1] and Yuyan Yan [1]

1 College of Geoscience and Surveying Engineering, China University of Mining and Technology-Beijing, Beijing 100083, China; ahao_wang@cumtb.edu.cn (A.W.)
2 Shanghai Astronomical Observatory, Chinese Academy of Sciences, Shanghai 200030, China
3 School of Astronomy and Space Science, University of Chinese Academy of Sciences, Beijing 100049, China
4 Shanghai Key Laboratory of Space Navigation and Positioning Techniques, Shanghai 200030, China
5 Chinese Academy of Surveying & Mapping, Beijing 100036, China
6 School of Environment Science and Spatial Informatics, China University of Mining and Technology, Xuzhou 221116, China; xuexiliu@cumt.edu.cn
* Correspondence: zhyize@shao.ac.cn

Abstract: Although the traditional Carrier-to-Code Leveling (CCL) method can provide ideal slant total electron content (STEC) observables for establishing ionospheric models, it must rely on dual-frequency (DF) receivers, which results in high hardware costs. In this study, an ionosphere-weight (IW) single-frequency (SF) precise point positioning (PPP) method for extracting STEC observables is proposed, and multi-global navigation satellite system (GNSS)-integrated processing is adopted to improve the spatial resolution of the ionospheric model. To investigate the advantages of this novel method, 41 European stations are used to establish the regional ionospheric model, and both low- and high-solar-activity conditions are considered. The results show that the IW SFPPP-derived regional ionospheric model has a significantly better quality of vertical total electron content (VTEC) than the CCL method when using the final global ionospheric map (GIM) as a reference, especially in areas with sparse monitoring stations. Compared with the CCL method, the RMS VTEC accuracy of the IW SFPPP method can be improved by 17.4% and 12.7% to 1.09 and 2.83 total electron content unit (TECU) in low- and high-solar-activity periods, respectively. Regarding GNSS carrier-phase-derived STEC variation (dSTEC) as the reference, the dSTEC accuracy of the IW SFPPP method is comparable to that of the CCL method, and its RMS values are about 1.5 and 2.8 TECU in low- and high-solar-activity conditions, respectively. This indicates that the proposed method using SF-only observations can achieve the same external accord accuracy as the CCL method in regional ionospheric modeling.

Keywords: multi-global navigation satellite system (multi-GNSS); precise point positioning (PPP); single frequency (SF); regional ionospheric modeling (RIM); ionosphere weighted (IW)

1. Introduction

The L-band radio signals broadcasted by the global navigation satellite system (GNSS) are subject to strong interference and cause uncertain path delays when passing through the ionosphere [1,2]. Although dual-frequency (DF) GNSS observations can eliminate

99.9% of ionospheric delay errors using an ionosphere-free (IF) combination model, single-frequency (SF) receivers or chips are still the preferred choice for GNSS users as they are more than ten times cheaper [3–5]. The positioning errors of standard point positioning (SPP) without ionospheric correction may exceed 20 m and can be three times worse than those of SPP based on the Neustrelitz total electron content model (NTCM) correction [6]. During periods of disturbances in solar and geomagnetic activity, the GNSS signal delay caused by ionospheric effects can even reach up to 100 m [7]. How to accurately handle ionospheric delay errors is the key to achieving high-precision positioning, and it is also an urgent problem to be solved in GNSS-related engineering applications.

Nowadays, there are two main types of ionospheric products: broadcast ionospheric model (BIM) and global ionospheric map (GIM). The BIM is widely used for real-time (RT) positioning since its coefficients can be obtained from the GNSS navigation message. Each GNSS has established its own BIM, such as the Klobuchar model for GPS, the NeQuick model for Galileo, and the BDGIM model for the Beidou global navigation satellite system (BDS-3). The ionospheric correction accuracy of BIM is relatively low, only about 50–75% on a worldwide scale, making it difficult to meet the requirements of high-precision positioning users [8–10]. As the most accurate global ionospheric product, GIM has the ability to support stable 1.5 m level SPP and 0.5 m level SF precise point positioning (PPP) services [5,11]. To meet both high-precision and RT requirements simultaneously, more and more Ionosphere Associate Analysis Centers (IAACs) are providing RT-GIM products for GNSS SF users. Some studies have confirmed that RT-GIM and post-processed GIM have comparable ionospheric correction accuracy [11,12]. Although the correction accuracy of (RT-)GIM can be as high as 2 total electron content unit (TECU), its quality may decrease to 8 TECU in areas with sparse ground-based GNSS stations or during ionospheric active periods [13]. In addition, satellite-based augmentation service (SBAS) and PPP-based real-time kinematic (PPP-RTK) technologies require extremely high accuracy for ionospheric correction. Therefore, high-quality regional ionospheric modeling plays an irreplaceable role in enhancing navigation.

The main factors affecting the quality of regional ionospheric modeling, under the premise of no change in the number and distribution of GNSS monitoring stations, include slant total electron content (STEC) extraction accuracy, vertical total electron content (VTEC) modeling algorithm, and ionospheric mapping function (MF) [14–16]. The STEC observables, as the data basis for ionospheric modeling, have a direct and significant impact on the quality of the ionospheric VTEC model. At present, the majority of ionospheric products released by IAACs use the traditional Carrier-to-Code Leveling (CCL) method to extract STEC observables [2,15]. This method adopts carrier-phase-smoothed code observations to avoid the problem of handling carrier-phase ambiguity and has the advantages of simple calculation and high processing efficiency. However, due to the limited accuracy of code observations and their susceptibility to code noise, multipath effects, and time-varying receiver DCB, the STEC observables extracted by using the CCL method cannot further improve the accuracy of ionospheric modeling [17–19]. To fully utilize the ultra-high-precision advantage of carrier-phase observations, the undifferenced and uncombined PPP (UU-PPP) method was proposed to extract the STEC observable [20]. Many contributions have proved that UU-PPP methods, including PPP-float or PPP-fixed solutions, are superior to the CCL method in ionospheric delay extraction [15,19,21–23].

Compared with the CCL method, which only adopts DF observations, the PPP method for extracting STEC observables is suitable not only for DF or multi-frequency (MF) observations but also for SF observations. Considering the current low price of SF GNSS receivers, the question of how to establish low-cost and high-precision regional ionospheric models using affordable SF monitoring equipment has become a hot topic in both atmosphere modeling and regional enhancing navigation. Considering that ionospheric delay causes the same numerical but opposite directional effects on code and carrier-phase observations, an ionospheric-free half-SFPPP approach was proposed to estimate the STEC values with the accuracy of submeters [24]. It is a pity that this study did not compare the performance of the observation method combined with the traditional CCL method. When applying UU-PPP technology to SF mode, even based on reparameterization, it is easy to cause rank deficiency in the normal equation. To accurately estimate the ionospheric parameters of each satellite in UU-SFPPP, the receiver clock needs to be reparametrized as the drift of the raw receiver clock relative to the first epoch. The root mean square (RMS) of the ionospheric VTEC differences between the UU-SFPPP and CCL methods is about 0.5 TECU [25]. In this study, we proposed a multi-GNSS ionosphere-weighted (IW) SFPPP method for extracting the ionospheric STEC observables. This novel method is still based on the UU-SFPPP model, but the problem of rank deficiency is solved by introducing the virtual ionospheric observations from the BIM. Meanwhile, the addition of external ionospheric constraints can enhance the model strength of UU-SFPPP in theory, thereby estimating more accurate STEC observables.

To investigate the advantages of the multi-GNSS IW-SFPPP in the STEC extraction and regional ionospheric modeling, the content of this paper is arranged as follows: first of all, the detailed algorithms for extracting the ionospheric STEC using the multi-GNSS IW-SFPPP method and establishing the regional ionospheric model are presented in Section 2. Then, the experimental datasets and processing strategies for regional ionospheric modeling are introduced in Section 3. In Section 4, referring to the GIM and GNSS geometry-free (GF) combinations, the quality of regional ionospheric models established by different STEC extraction methods is evaluated and compared. Finally, some remarkable conclusions and new findings are summarized in Section 5.

2. Methodology

With the comprehensive operation of the BDS-3 and Galileo systems, more and more GNSS satellites can be used in ionospheric modeling, which can effectively improve the spatial resolution of the ionosphere pierce point (IPP) without modifying the ground-based monitoring networks. Thus, GPS, Galileo, and BDS-3 satellites are simultaneously applied in the proposed IW SFPPP method to obtain high-spatial–temporal-resolution ionospheric STEC observables. In addition, the modeling algorithms for regional ionospheric VTEC are described in this section.

2.1. General Model of the GNSS Single-Frequency PPP

The original GNSS observations of code $P_{r,i}^{s,Q}$ and carrier-phase $L_{r,i}^{s,Q}$ at the first frequency can be expressed as [23,26]

$$\begin{cases} P_{r,1}^{s,Q} = \Delta_{r,1}^{s,Q} + c \cdot (dt_r - dt^{s,Q}) + T_r + I_1^{s,Q} + B_{r,1} - B_1^{s,Q} + \varepsilon_p \\ L_{r,1}^{s,Q} = \Delta_{r,1}^{s,Q} + c \cdot (dt_r - dt^{s,Q}) + T_r - I_1^{s,Q} + N_1^{s,Q} + b_{r,1} - b_1^{s,Q} + \varepsilon_L \end{cases} \quad (1)$$

where the superscript s denotes the GNSS satellite and Q represents the specific GNSS. The subscript r denotes the GNSS receiver. c denotes the light velocity. $\Delta_{r,1}^{s,Q}$ denotes the geometric distance between the GNSS satellite and the receiver. $dt^{s,Q}$ and dt_r denote the satellite and receiver clock offsets, respectively. T_r denotes the tropospheric delay error in the line-of-sight direction. $I_1^{s,Q}$ denotes the ionospheric delay error in the line-of-sight direction. $N_1^{s,Q}$ denotes the integer ambiguity of the carrier-phase observation. $B_{r,1}$ and $B_1^{s,Q}$ denote the code hardware delays for the receiver and satellite. $b_{r,1}$ and $b_1^{s,Q}$ denote the carrier-phase hardware delays for the receiver and satellite. ε_p and ε_L denote the code and carrier-phase observation noises, including multipath effects and unmodeled errors.

Considering that the precise ephemeris based on the IF combinations cannot be directly applied to the SF observation, the satellite clock offset $dt^{s,Q}$ in Equation (1) needs to be recalculated through differential code bias (DCB) correction [27]. The new clock offsets for the satellite $d\bar{t}^{s,Q}$ and receiver $d\bar{t}_r$ can be expressed as

$$\begin{cases} d\bar{t}^{s,Q} = dt^{s,Q} + \frac{d_{IF}^{s,Q}}{c} \\ d\bar{t}_r = dt_r + \frac{d_{r,IF}}{c} \end{cases} \tag{2}$$

with

$$\begin{cases} d_{IF}^{s,Q} = \frac{f_1^2 \cdot B_1^s - f_i^2 \cdot B_i^s}{f_1^2 - f_i^2} \\ d_{r,IF} = \frac{f_1^2 \cdot B_{r,1} - f_i^2 \cdot B_{r,i}}{f_1^2 - f_i^2} \\ DCB^s = B_1^s - B_i^s \\ DCB_r = B_{r,1} - B_{r,i} \end{cases} \tag{3}$$

where $d_{IF}^{s,Q}$ and $d_{r,IF}$ denote the IF combined code hardware delay for the satellite and receiver. f denotes the GNSS signal frequency. The subscript i represents the frequency number; $i = 2$ and $i = 3$ for the GPS/Galileo and BDS-3 satellites, respectively. DCB^s and DCB_r denote the satellite and receiver DCB corrections.

Substituting Equation (2) and (3) into Equation (1), the new code and carrier-phase observations of one satellite can be expressed as

$$\begin{cases} P_{r,i}^{s,Q} = \Delta_{r,i}^{s,Q} + c \cdot (d\bar{t}_r - d\bar{t}^{s,Q}) + T_r + I_1^{s,Q} - \frac{f_i^2}{f_1^2 - f_i^2}(DCB_r - DCB^s) + \varepsilon_P \\ L_{r,i}^{s,Q} = \Delta_{r,i}^{s,Q} + c \cdot (d\bar{t}_r - d\bar{t}^{s,Q}) + T_r - I_1^{s,Q} + N_1^{s,Q} + d_{r,IF} - d_{IF}^s + b_{r,1} - b_1^{s,Q} + \varepsilon_L \end{cases} \tag{4}$$

In SFPPP, the satellite orbits and clocks can be fixed using precise ephemeris. The tropospheric delay errors are first corrected using the tropospheric empirical or numerical models, and then parameter estimation can be used to eliminate the uncorrectable tropospheric wet components [28,29]. Considering that the code/carrier-phase hardware delays and ionosphere/ambiguity parameters have a strong correlation, it is necessary to reparametrize these estimable parameters. Equation (4) can be re-written as

$$\begin{cases} P_{r,i}^{s,Q} = A_r^{s,Q} \cdot \Omega + e_t \cdot d\bar{t}_r + M_{tro} \cdot T_{zwd} + \bar{I}_1^{s,Q} + \varepsilon_P \\ L_{r,i}^{s,Q} = A_r^{s,Q} \cdot \Omega + e_t \cdot d\bar{t}_r + M_{tro} \cdot T_{zwd} - \bar{I}_1^{s,Q} + \bar{N}_1^{s,Q} + \varepsilon_L \end{cases} \tag{5}$$

with

$$\begin{cases} \bar{I}_1^{s,Q} = I_1^{s,Q} - \frac{f_i^2}{f_1^2 - f_i^2}(DCB_r - DCB^s) \\ \bar{N}_1^{s,Q} = N_1^{s,Q} + d_{r,IF} - d_{IF}^s + b_{r,1} - b_1^{s,Q} - \frac{f_i^2}{f_1^2 - f_i^2}(DCB_r - DCB^s) \end{cases} \tag{6}$$

where Ω and A denote the vector of the positioning errors and its unit vector of the coordinate component, respectively. $d\bar{t}_r$ denotes the new receiver clock offset that absorbs the code hardware delay, and e_t is the corresponding unit vector. T_{zwd} denotes the residual error of the tropospheric delay. M_{tro} denotes the tropospheric mapping function. $\bar{I}_1^{s,Q}$ denotes the new ionospheric delay parameter that includes both satellite and receiver DCB. It should be noted that the new ambiguity of carrier-phase $\bar{N}_1^{s,Q}$ has lost the integer property due to absorbing both code and carrier-phase hardware delays.

2.2. Extraction of STEC Observables from the Multi-GNSS Ionosphere-Weighted Single-Frequency PPP

Although the number of estimated parameters in Equation (5) can be reduced by adopting reparameterization, it still results in a rank deficiency of normal equation in the case of estimating ionospheric parameters for each satellite. This problem can be solved by introducing virtual ionospheric observations based on the BIM. It is worth noting that the ionospheric delay errors $\overline{I}_1^{s,Q}$ in Equation (5) can be corrected with BIM in advance and its residual error as a new parameter to be estimated in IW SF-PPP. To further extract more accurate ionospheric STEC observables, the positioning errors of the monitoring station should not be estimated but fixed using precise coordinates. Thus, the positioning errors Ω in Equation (5) need to be removed and the modified equations for extracting STEC observables can be expressed as

$$\begin{cases} P_{r,i}^{s,G} = e_t \cdot \overline{dt}_r^G + M_{tro} \cdot T_{zwd} + \overline{I}_1^{s,G} + \varepsilon_P^G \\ P_{r,i}^{s,E} = e_t \cdot \overline{dt}_r^G + \delta_{ISB}^E + M_{tro} \cdot T_{zwd} + \overline{I}_1^{s,E} + \varepsilon_P^E \\ P_{r,i}^{s,C} = e_t \cdot \overline{dt}_r^G + \delta_{ISB}^C + M_{tro} \cdot T_{zwd} + \overline{I}_1^{s,C} + \varepsilon_P^C \\ L_{r,i}^{s,G} = e_t \cdot \overline{dt}_r^G + M_{tro} \cdot T_{zwd} - \overline{I}_1^{s,G} + \overline{N}_1^{s,G} + \varepsilon_L^G \\ L_{r,i}^{s,E} = e_t \cdot \overline{dt}_r^E + \delta_{ISB}^E + M_{tro} \cdot T_{zwd} - \overline{I}_1^{s,E} + \overline{N}_1^{s,E} + \varepsilon_L^E \\ L_{r,i}^{s,C} = e_t \cdot \overline{dt}_r^C + \delta_{ISB}^C + M_{tro} \cdot T_{zwd} - \overline{I}_1^{s,C} + \overline{N}_1^{s,C} + \varepsilon_L^C \\ v_{iono}^{s,G/E/C} = k_1^{s,G/E/C} + \varepsilon_{iono}^{G/E/C} \end{cases} \quad (7)$$

where $\overline{dt}_r^G$ is the designated receiver clock offset; thus, inter-system bias (ISB) δ_{ISB}^E and δ_{ISB}^C need to be set for Galileo and BDS-3 satellites. v_{iono}^s denotes the virtual observations of the ionospheric delay for each satellite, and ε_{iono} denotes the ionospheric noise. It should be emphasized that k_1^s can be obtained from the BIM product. In this contribution, the classical GPS Klobuchar model is selected for constructing the virtual ionospheric observations, because it is suitable for RT processing and has high computational efficiency. It should be noted that the accuracy of the external ionospheric model mainly affects the convergence speed, with little impact on positioning accuracy. Thus, the GPS Klobuchar as external ionospheric constraint can fully meet the accuracy requirements of the IW SFPPP [5]. Before estimating ionospheric residuals, it is necessary to set appropriate a priori variance for ionospheric parameters, and the square of the Klobuchar correction value can be used for ionospheric weighting in IW SFPPP [5].

In the multi-GNSS IW SFPPP, the estimable parameters can be summarized as $[\overline{dt}_r^G, \delta_{ISB}^E, \delta_{ISB}^C, T_{zwd}, \overline{I}_1^{s,G}, \overline{I}_1^{s,E}, \overline{I}_1^{s,C}, \overline{N}_1^{s,G}, \overline{N}_1^{s,E}, \overline{N}_1^C]$. Thus, the ionospheric STEC in unit TECU can be expressed as

$$STEC = \overline{I}_1^{s,Q} \cdot \frac{(f_1)^2}{40.3 \times 10^{16}} \quad (8)$$

2.3. Algorithm of the Regional Ionospheric VTEC Model

The single-layer model is applied to construct the regional ionospheric VTEC model. The ionospheric VTEC can be obtained from Equation (8) through the mapping function conversion as follows [30]:

$$\begin{cases} VTEC = \Phi_{mf} \cdot STEC \\ \Phi_{mf} = \cos(\arcsin(\frac{R_e}{R_e + H_{ion}} \cdot \sin(Z_r^s))) \end{cases} \quad (9)$$

where Φ_{mf} is the ionospheric mapping function. Z_r^s denotes the zenith distance at the monitoring GNSS stations. $R_e = 6371$ km denotes the mean radius of the Earth. $H_{ion} = 450$ km denotes the height of the single layer and is consistent with the value adopted by the Center for Orbit Determination in Europe (CODE) agency.

By merging Equations (6), (8) and (9), the observation equation of the ionospheric modeling based on the IW SFPPP method can be summarized as

$$\frac{\Phi_{mf} \cdot f_1^2}{40.3 \times 10^{16}} \cdot \overline{I}_1^{s,Q} = VTEC + \frac{\Phi_{mf} \cdot f_1^2 \cdot f_i^2}{40.3 \times 10^{16} \cdot (f_1^2 - f_i^2)} \cdot (DCB_r - DCB^s) \tag{10}$$

Due to the difficulty in meeting the orthogonality of spherical harmonic models in regional ionospheric modeling, the polynomial model is applied to model the regional VTEC as follows [31]:

$$VTEC = \sum_{i=0}^{n} \sum_{j=0}^{m} E_{ij}(\lambda - \lambda_0)^i (\beta - \beta_0)^j \tag{11}$$

where E_{ij} denote estimable coefficients of the regional ionospheric VTEC model. λ and β denote the latitude and longitude of the IPP. λ_0 and β_0 denote the latitude and longitude of the geometric center in the region. n and m denote the max order and degree of the polynomial function.

From Equation (10), it can be seen that the ionospheric observables extracted by the IW SFPPP method can be affected by DCB errors. Since the satellite DCB obtained from BSX (Bias Solution Independent Exchange Format) products has an accuracy better than 0.1 ns, it can be fixed using CAS (Chinese Academy of Sciences) daily solutions [27]. However, it is difficult to precisely correct the receiver DCB with BSX corrections due to its distinct short-term time-varying property [32]. Hence, the receiver DCB needs to be estimated together with the ionospheric model parameters.

3. Experimental Data and Processing Strategies

To assess the quality of the proposed ionospheric extraction method, a mid-latitude region in the Northern Hemisphere was selected to establish the regional ionospheric VTEC model. In addition, some key processing strategies involved in IW SFPPP are summarized in this section.

3.1. Experimental Data

Forty-one multi-GNSS experiment (MGEX) stations located in Europe are used to establish the regional ionospheric VTEC model. This experimental area covers a northern latitude of 35–65 degrees and an eastern longitude of 0–30 degrees. The distribution of all selected stations marked in red is displayed in Figure 1a. Twelve well-distributed blue stations (as shown in Figure 1b) in this monitoring network are used for testing the IW SFPPP performance and evaluating the external accord accuracy of the regional ionospheric VTEC model.

When selecting the ionospheric modeling time, we considered different solar and geomagnetic activity during solar cycle 25. Figure 2 shows the solar radio flux at 10.7 cm wavelength (download link: https://wdc.kugi.kyoto-u.ac.jp/kp/index.html#LIST (accessed on 8 February 2025)) during the two periods. The day of year (DoY) 153–162 in 2021 corresponds to the quiet period with the F10.7 index between 73.3 and 80.8 sfu (solar flux units), while in 2024, its F10.7 index varies from 177.8 to 226.9 sfu. This indicates that the solar activity in 2024 is particularly active. On the other hand, the geomagnetic Kp index (download link: https://wdc.kugi.kyoto-u.ac.jp/wdc/Sec3.html (accessed on 8 February 2025)) during the same period is shown in Figure 3. Most Kp index values in 2021 can be kept within 1–2 and no more than 3.67, indicating calm geomagnetic activity. In contrast, the Kp index in 2024 is higher, with several peaks exceeding 5, reflecting the active state of the ionosphere during this period.

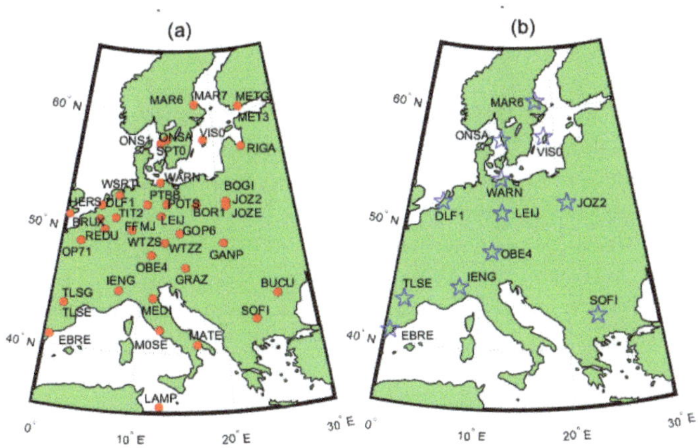

Figure 1. Distribution of the selected GNSS monitoring stations (41 red stations (**a**), 12 blue stations (**b**)).

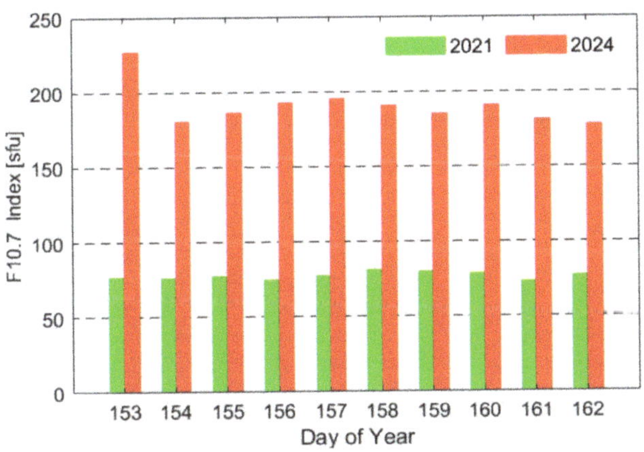

Figure 2. Solar radio flux at 10.7 cm wavelength during the testing period.

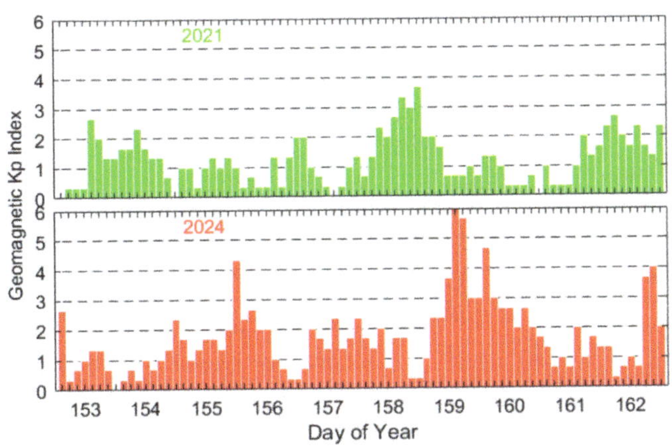

Figure 3. Geomagnetic Kp index during the testing period.

3.2. Processing Strategies

In regional ionospheric VTEC modeling based on the IW SFPPP method, the adopted SF observations are C1W/L1C for GPS, C1C/L1C and C1X/L1X for Galileo, and C2I/L2I for BDS-3. For the traditional CCL method, the C1W/L1C+C2W/L2W observations of GPS, C1C/L1C+C5Q/L5Q and C1X/L1X+C5X/L5X observations of Galileo, and C2I/L2I+C6I/L6I observations of BDS-3 are used. In the process of both IW SFPPP and regional ionospheric modeling, the elevation cutoff angle is set to 10 degrees. The priori precision of the code and carrier-phase observations for GPS/Galileo/BDS-3 is set to 0.3 and 0.003 m, respectively, and the elevation-dependent model is used to weight observations of each satellite [33]. The GBM precise ephemeris provided by the Deutsches GeoForschungsZentrum (GFZ) was used to fix the satellite orbits and clocks. The precise coordinates of the MGEX stations used are fixed using the SINEX (Solution-INdependent EXchange) products. The relativistic effects, solid/ocean/polar tidal errors, and phase windup are corrected with traditional models [34,35]. In addition, the main processing strategies and adopted models are summarized in Table 1.

Table 1. Processing strategies of the multi-GNSS IW SFPPP and regional ionospheric modeling.

Items	Strategies
I: IW SFPPP processing	
Estimator	Kalman filter
Satellite/receiver antenna center offsets and variations	Corrected with igs20_2317.atx products
Satellite DCB	Corrected with CAS daily BSX products
Tropospheric delay	Dry delay: corrected with GPT2w+SAAS+VMF models Wet delay: estimated as random-walk process [36]
Ionospheric delay	Estimated as random-walk process [37]
Receiver clock	Estimated as white noise
Galileo and BDS-3 ISB	Estimated as random-walk process [38]
Phase ambiguity	Estimated as float solution
II: Regional ionospheric VTEC modeling	
Estimator	Sequential least-squares adjustment
VTEC modeling algorithm	Polynomial function (6 orders $\times$ 4 degrees, 2 h interval)
Receiver DCB	Estimated as constant

4. Results and Discussion

The static positioning performance of GPS/Galileo/BDS-3 integrated IW SFPPP needs to be investigated. It can indirectly reflect the feasibility of the proposed ionospheric extraction method. To objectively evaluate the advantages of this novel method, the ionospheric model established by the IW SFPPP method was compared with the CODE GIM product and the CCL-based model. In addition, the external accord accuracy of the new method was also evaluated using the epoch-differenced GF combinations as reference.

4.1. Performance of the Multi-GNSS Ionosphere-Weighted Single-Frequency PPP

The twelve blue stations shown in Figure 1b were selected for conducting the multi-GNSS IW SFPPP in static mode. The RMS values of positioning errors for all stations within 10 days are displayed in Figure 4. The abbreviations 'Hor' and 'Ver' in this figure denote the horizontal and vertical components. Both horizontal and vertical positioning errors of the GPS+Galileo+BDS-3 IW SFPPP in 2021 and 2024 can be less than 10 cm within about 70 min. After converging to 10 cm, the horizontal positioning errors in 2021 are significantly larger than those in 2024, mainly due to the smaller number of available BDS-3 satellites in 2021. In 2024, the positioning errors of both horizontal and vertical components can stabilize at 5 cm after 150 min.

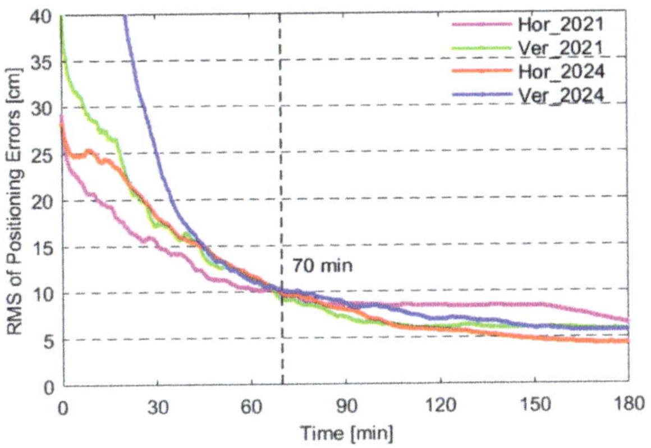

Figure 4. RMS of positioning errors of the static IW SFPPP during the first 3 h (12 blue stations, DoY 153–162, 2021 and 2024).

Figure 5 gives the RMS positioning accuracy of the static multi-GNSS IW SFPPP after 3 h convergence for all tested stations. The positioning accuracy of most stations in 2024 is better than that in 2021, especially in the vertical direction. The mean RMS errors of all stations in 2021 can reach 1.6, 2.6, and 6.3 cm in the north (N), east (E), and up (U) directions, respectively. By comparison, the vertical positioning accuracy in 2024 has improved from 6.3 to 4.5 cm. This indicates that the current SFPPP through the multi-GNSS fusion and IW constraints can achieve high-precision positioning results, with horizontal results better than 3 cm and vertical results better than 5 cm. Therefore, it can be proven that the multi-GNSS IW SFPPP method for extracting ionospheric observables is theoretically feasible. More importantly, the monitoring station coordinates are fixed using the SINEX solutions instead of being estimated as parameters, which will further improve the accuracy of extracting ionospheric observables.

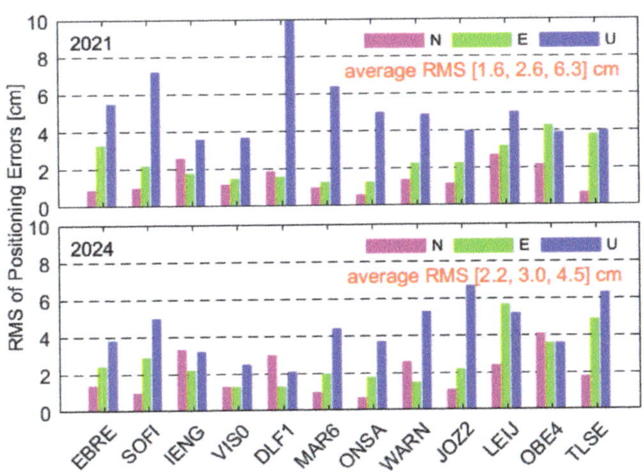

Figure 5. RMS positioning accuracy of the IW SFPPP in static mode (DoY 153–162, 2021 and 2024).

4.2. *Quality Assessment of the Regional Ionospheric VTEC Model in Comparison with GIM*

In this study, 35 coefficients (6 orders and 4 degrees in polynomial function) of the regional ionospheric VTEC model are estimated with an interval of 2 h. Figure 6 shows the gridded ionospheric VTEC maps with the spatial resolution of $0.1° \times 0.1°$ on DoY 157, 2021, which are calculated from the IW SFPPP-derived slant ionospheric observables. We can see that the daily VTEC values are no more than 25 TECU, and some values can even be less than 5 TECU at GPS time (GPST) 00:00 and 02:00. This indicates that the ionospheric VTEC values in low-solar-activity years (seen in Figure 2) are also relatively small. There are significant differences in VTEC values at different latitudes, with VTEC in higher-latitude areas generally smaller than those in lower-latitude areas. The same processing is applied to the regional ionospheric VTEC maps on DoY 157, 2024, and the corresponding results are given in Figure 7. Due to the high solar activity in 2024, the peak of VTEC values in daily solutions can reach up to 60 TECU and its minimum value exceeds 10 TECU. Under this ionospheric condition, the differences in ionospheric VTEC between different latitude bands are quite distinct.

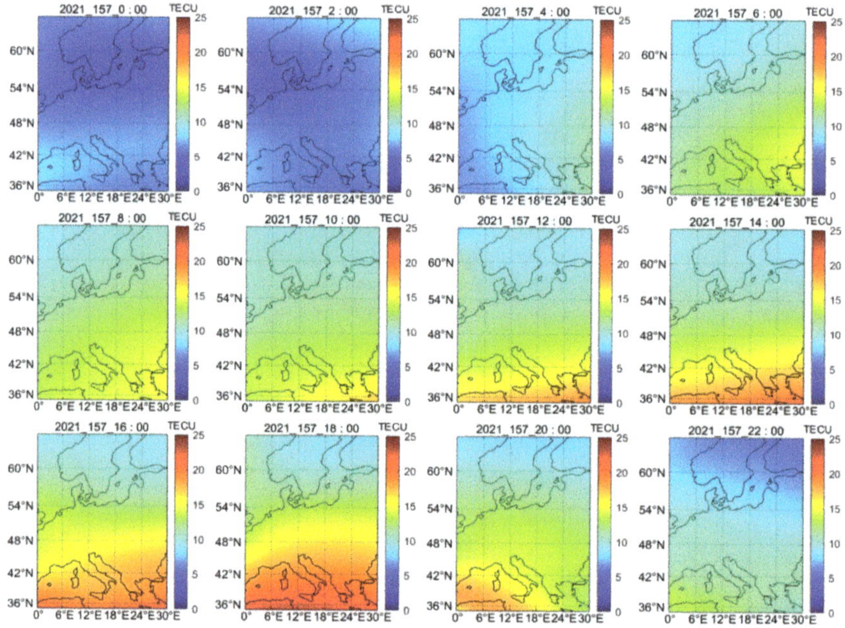

Figure 6. Regional ionospheric VTEC maps based on the IW SFPPP-derived STECs on DoY 157, 2021.

Figure 8 summarizes the RMS of ionospheric VTEC values derived from the IW SFPPP method. The error bars in this figure represent the STD of ionospheric VTEC values. The ionospheric VTEC values in 2024 with about 30 TECU are much larger than those in 2021 with about 12 TECU, which is consistent with the trend of the F10.7 index during the same period (seen in Figure 2). From the results of the error bars, it can be seen that the mean STD of ionospheric VTEC in 2024 can reach 8 TECU, while in 2021, the corresponding statistical value is around 3.7 TECU. This proves that the ionospheric VTEC in 2024 have greater variation amplitude in daily solutions.

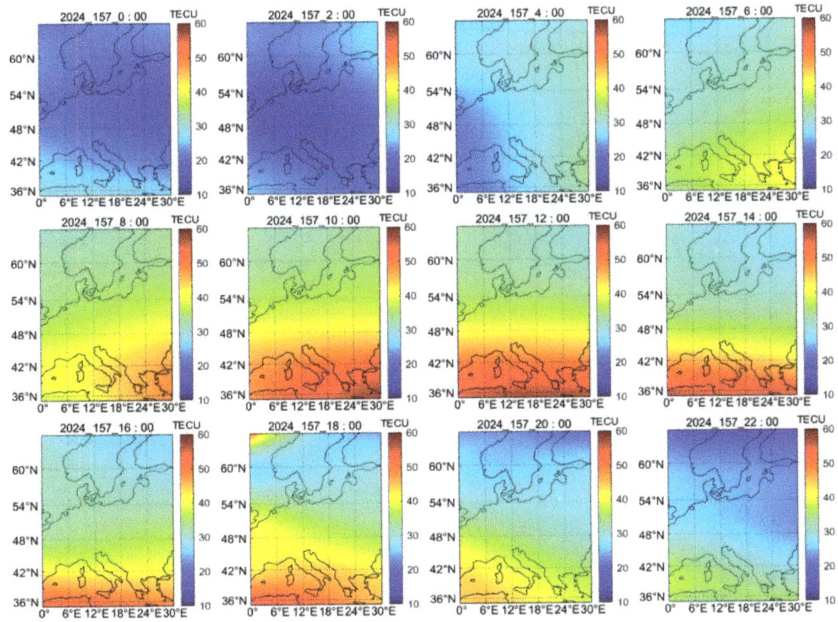

Figure 7. Regional ionospheric VTEC maps based on the IW SFPPP-derived STECs on DoY 157, 2024.

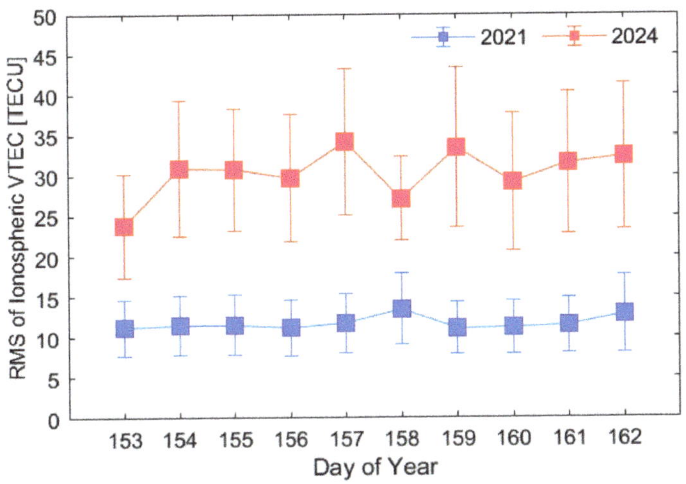

Figure 8. RMS of the regional ionospheric VTEC values based on the IW SF-PPP-derived STEC.

In this section, the post-processing GIM products provided by CODE are used to judge the quality of the regional ionospheric VTEC models. The comparison interval is set to 2 h, and every day has 12 groups of VTEC differences (i.e., dVTEC) between the regional ionospheric VTEC and final GIM products. Figures 9 and 10 show the RMS maps of dVTEC on DoY 153-162 in 2021 derived from the IW SFPPP and CCL methods, respectively. It can be seen that the RMS values of most dVTEC for the IW SFPPP method are better than 1 TECU and have good consistency in spatial distribution. For the CCL method, the RMS peak of dVTEC on DoY 153-155 and 160 may reach up to 5 TECU, which may be attributed to the sparsely distributed monitoring stations in these areas. By comparing

Figures 9 and 10, we can see that using the IW SFPPP method to construct the regional ionospheric VTEC model can achieve better performance in areas with sparse stations.

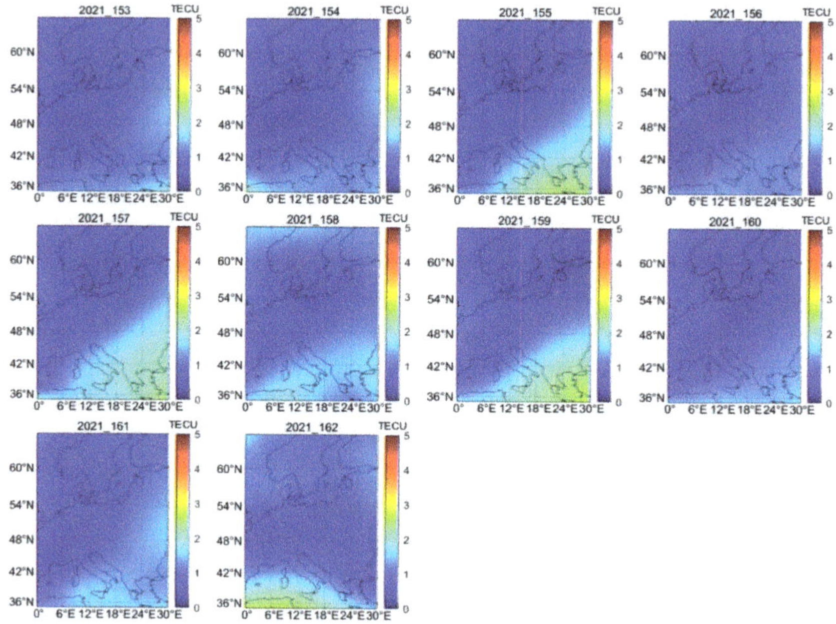

Figure 9. RMS maps of VTEC differences between the IW SFPPP method and final GIM products on DoY 153–162 in 2021.

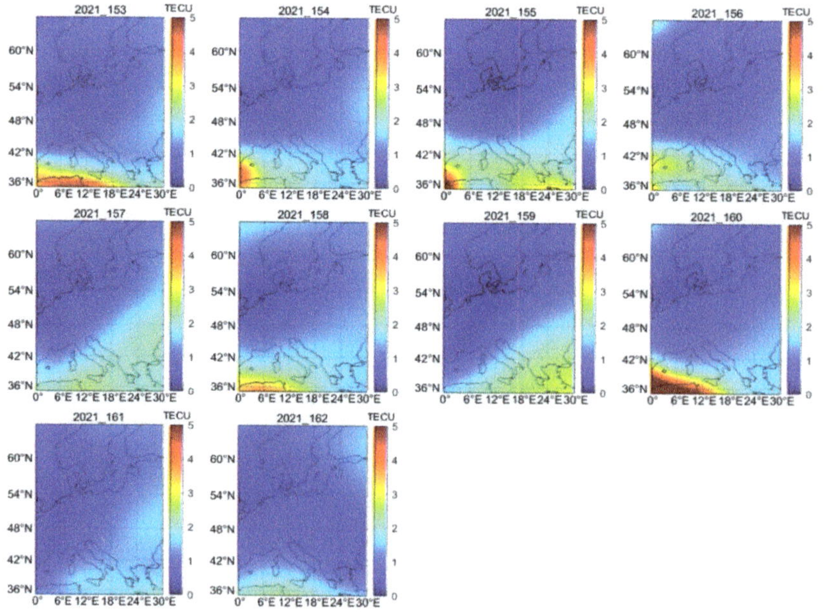

Figure 10. RMS maps of VTEC differences between the CCL method and final GIM products on DoY 153–162 in 2021.

The same comparative strategies are applied to the high-solar-activity year of 2024. The corresponding RMS maps of dVTEC generated by the IW SFPPP and CCL methods are presented in Figures 11 and 12, respectively. The variation amplitude of the daily dVTEC solutions in 2024 is more intense than that in 2021, with peak values of up to 8 TECU, which is reasonable given that ionospheric VTEC is more difficult to model under high solar and geomagnetic conditions. The majority of dVTEC values of the IW SFPPP method are significantly lower than those of the CCL method on DoY 155–156 and 159–160. More than half of dVTEC for the IW SFPPP can be less than 3 TECU in daily solutions, while for the CCL method, the corresponding dVTEC values are kept within 3 to 5 TECU. This proves that the IW SFPPP method still outperforms the CCL method in high-solar-activity years.

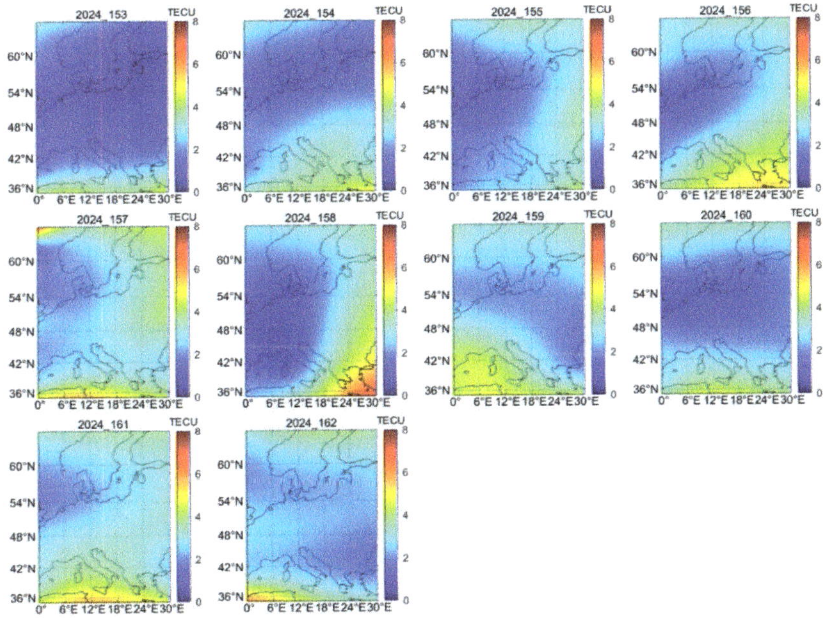

Figure 11. RMS maps of VTEC differences between the IW SFPPP method and final GIM products in 2024.

Figure 13 summarizes the RMS values of VTEC differences between the regional ionospheric model and final GIM products on different days. Compared with the CCL method with a mean RMS of 1.32 TECU in 2021, the corresponding accuracy of dVTEC based on the IW SFPPP method can be improved by 17.4% to 1.09 TECU. In the high-solar-activity year of 2024, the modeling precision of the regional ionospheric VTEC sharply decreased, and its mean RMS values of dVTEC are reduced to 2.83 and 3.24 TECU for the IW SFPPP and CCL methods, respectively. The improvement in dVTEC accuracy using the IW SFPPP method is about 12.7% compared with the CCL method. No matter what the solar and geomagnetic activity conditions are, the conformity between the regional ionospheric VTEC modeled by the IW SFPPP method and final GIM products is significantly better than that of the CCL method.

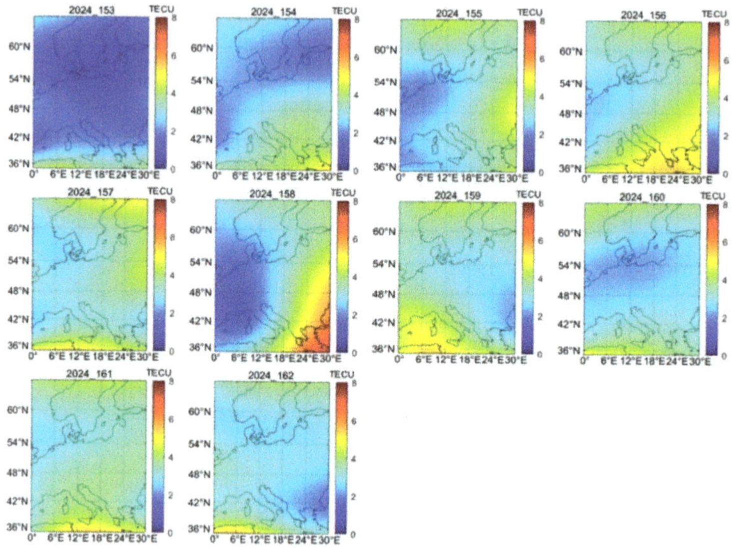

Figure 12. RMS maps of VTEC differences between the CCL method and final GIM products in 2024.

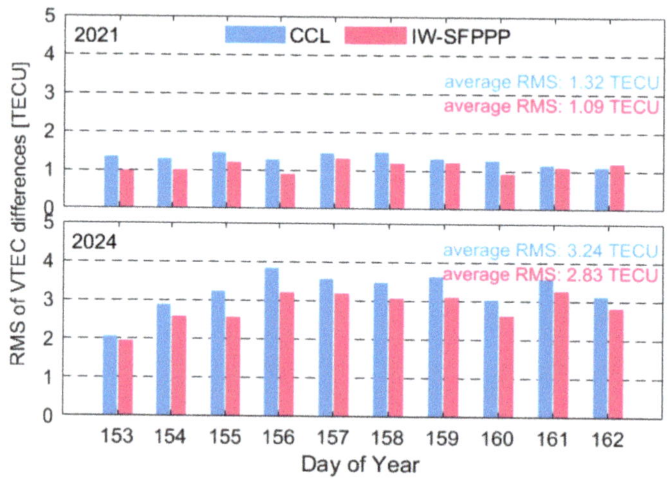

Figure 13. RMS of VTEC differences between the regional ionospheric model and final GIM products.

In order to more intuitively reflect the differences between the IW SFPPP and CCL methods in regional ionospheric VTEC modeling, the RMS maps of dVTEC derived from the two abovementioned methods in 2021 and 2024 are shown in Figures 14 and 15, respectively. Most dVTEC values in 2021 are under 1 TECU, while in 2024, the corresponding values vary between 1 and 2 TECU. This indicates that the difference in modeling accuracy between the IW SFPPP and CCL methods may be increased as solar activity strengthens. In both low- and high-solar-activity years, the larger differences in dVTEC are presented in areas with sparse monitoring stations and their maximum values can be up to 3 TECU. The main reason for this phenomenon may be that the CCL method has fewer IPPs in sparse areas after removing gross errors. The IW SFPPP method with a large number of IPPs can demonstrate superior performance in areas with sparse monitoring stations. Figure 16 gives the RMS of VTEC differences between the IW SFPPP and CCL methods on each test day. We can see that the mean RMS of dVTEC values for all test days is 0.72 and 1.06 TECU in 2021 and 2024, respectively. This illustrates the conformity of VTEC for the two methods in 2021 is better than that in 2024, as the ionospheric VTEC can be more precisely modeled in calm solar and geomagnetic activity.

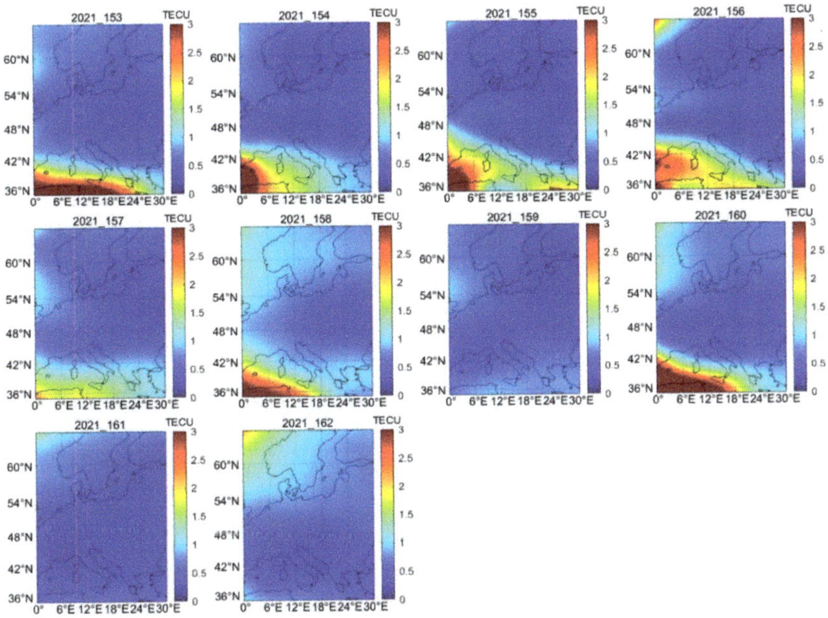

Figure 14. RMS maps of VTEC differences between the IW SFPPP method and CCL method in 2021.

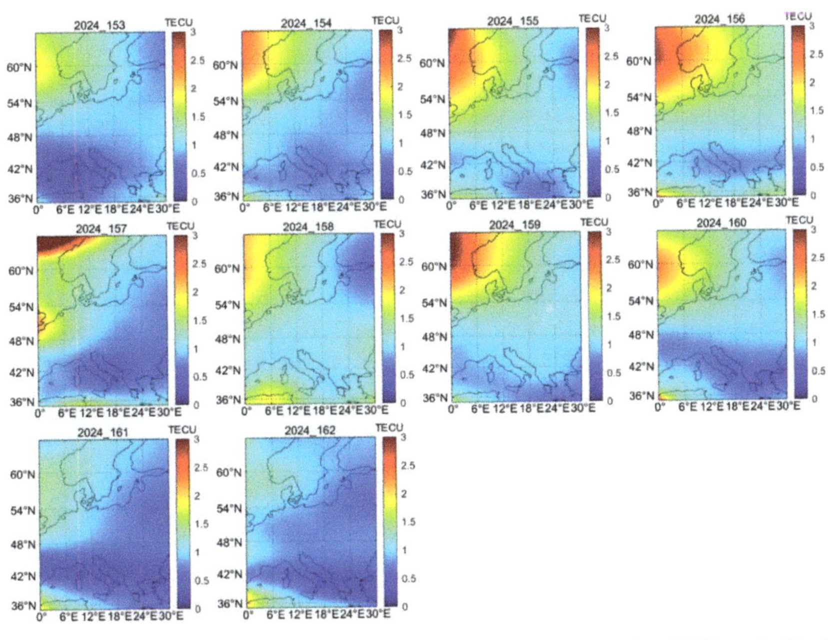

Figure 15. RMS maps of VTEC differences between the IW SFPPP method and CCL method in 2024.

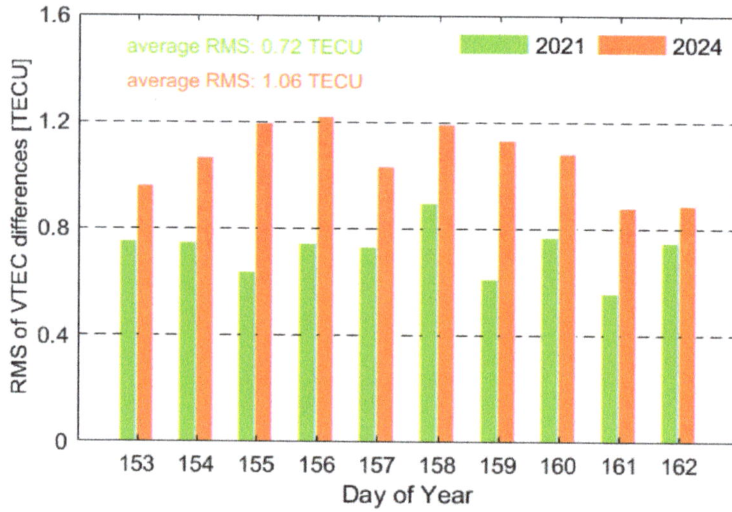

Figure 16. RMS of VTEC differences between the IW SFPPP method and CCL method.

4.3. Evaluation of External Accord Accuracy for Regional Ionospheric VTEC Model

Considering that the epoch-differenced GF combinations of carrier-phase observations have the ability to obtain the mm level STEC variations (i.e., dSTEC), the dSTEC can be adopted as the referenced values to verify the external accord accuracy of different ionospheric VTEC models [39]. The dSTEC modeled values corresponding to the above dSTEC referenced values are calculated from different regional ionospheric VTEC models (IW SFPPP and CCL methods) and CODE final GIM products. Thus, the dSTEC differences can be defined as the dSTEC modeled values minus the dSTEC reference values. In this section, twelve well-distributed stations marked in blue, as shown in Figure 1b, are used to implement the dSTEC comparison. The sampling rate of dSTEC is set to 5 min. Figures 17 and 18 show the dSTEC differences in four representative stations on DoY 157 in 2021 and 2024, respectively. It can be seen that dSTEC differences in 2024 are significantly larger than those in 2021 for both regional and global ionospheric VTEC models, which is caused by the larger ionospheric delays and drastic changes during high-solar-activity years. As the satellite elevation increases, the dSTEC differences sharply decrease. When the satellite elevation is higher than 40 degrees, the dSTEC differences in 2021 can be maintained within 4 TECU, but in 2024, the corresponding value is increased to 6 TECU. On the other hand, the distribution of dSTEC differences for both IW SFPPP and CCL methods is not as concentrated as for GIM products. This indicates that the quality of the final GIM products is slightly better than that of the regional ionospheric VTEC models established in this contribution.

Figure 19 gives the RMS of dSTEC differences for the IW SFPPP, CCL, and GIM models on DoY 153–162 in 2021 and 2024. The mean RMS values of dSTEC differences in 2021 are 1.54, 1.56, and 1.48 TECU for the IW SFPPP method, CCL method, and GIM products, respectively. In the high-solar-activity year of 2024, the corresponding values are increased to 2.82, 2.80, and 2.61 TECU, respectively. By evaluating the external accord accuracy of different ionospheric VTEC models, the quality of the regional ionospheric VTEC model established by the IW SFPPP method can be fully comparable to that of the CCL method, and its accuracy difference with the final GIM products can be maintained within 0.2 TECU.

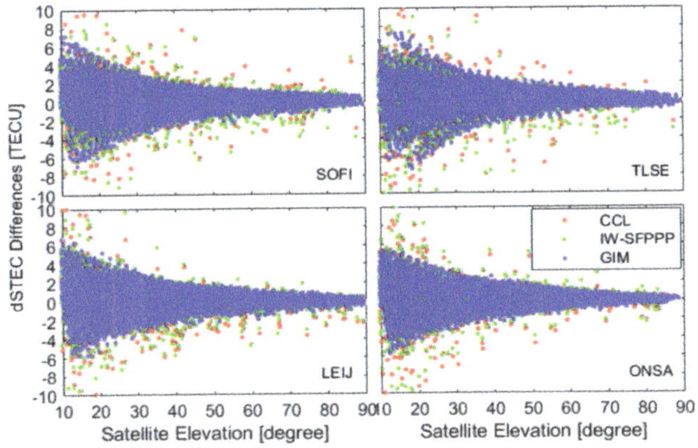

Figure 17. dSTEC differences for the IW SFPPP method, CCL method, and GIM products on DoY 157 in 2021.

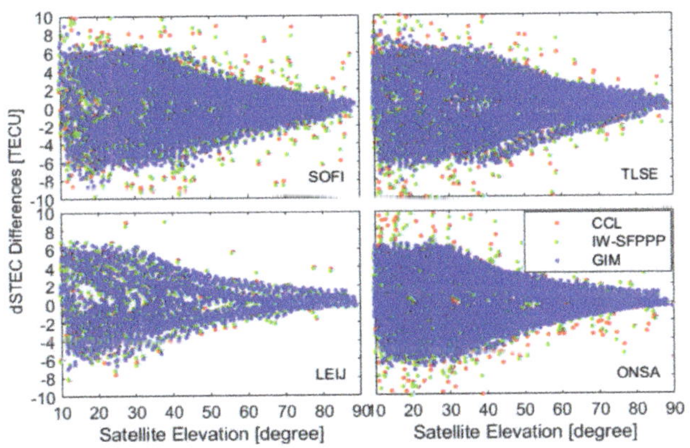

Figure 18. dSTEC differences for the IW SFPPP method, CCL method, and GIM products on DoY 157 in 2024.

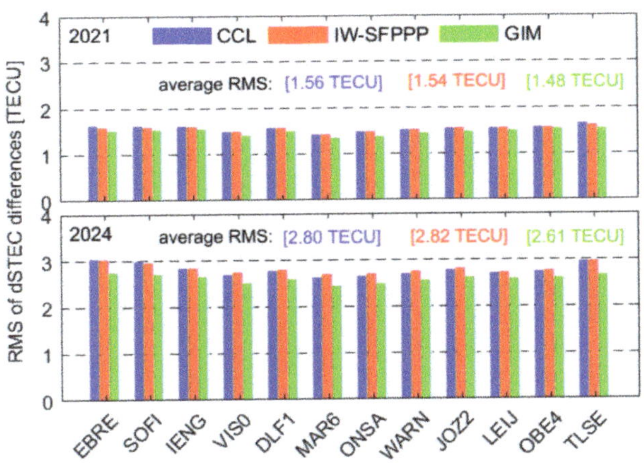

Figure 19. RMS of dSTEC differences for the IW SFPPP, CCL and GIM models at 12 blue stations (DoY 153–162, 2021 and 2024).

5. Conclusions

The classical CCL method, a commonly used technique for extracting ionospheric observables, has ideal performance and reliability. However, it must rely on DF receivers, which results in high hardware costs for ionospheric modeling. Considering that current UU-SFPPP can achieve a positioning accuracy that is better than 5 cm with the combination of multi-GNSS and external ionospheric constraints, the multi-GNSS IW SFPPP can be expected to become a high-precision and low-cost method for extracting ionospheric observables. In order to investigate the advantages of this novel STEC extraction method in regional ionospheric modeling, forty-one MGEX stations located in Europe are used to establish ionospheric models, and both low- and high-solar-activity conditions are considered.

Using the VTEC provided by the final CODE GIM as a reference, the regional ionospheric model based on the IW SFPPP method has a significantly better VTEC accuracy than the CCL method. In the low-solar-activity year of 2021, compared with the CCL method, the RMS VTEC accuracy of the IW SFPPP method can be improved by 17.4% to 1.09 TECU, while in the high-solar-activity year of 2024, the corresponding improvement ratio is about 12.7% for the IW SFPPP method. From the daily solutions of the VTEC difference between the regional ionospheric model and final GIM products, the modeling accuracy of the IW SFP method in areas with sparse monitoring stations is better than that of the CCL method. In addition, in the direct comparison between the IW SFPPP and CCL methods, the RMS of VTEC differences in 2021 and 2024 is about 0.72 and 1.06 TECU, respectively.

In the external accord accuracy evaluation of the regional ionospheric models, the GNSS dSTEC with mm level accuracy is adopted as the reference value. Regardless of the strength of solar activity, the accuracy of dSTEC differences for the IW SFPPP method is comparable to that of the CCL method, and its RMS values are about 1.5 and 2.8 TECU in 2021 and 2024, respectively. This indicates that the proposed IW SFPPP method using only SF observations can achieve the same ionospheric modeling accuracy as the CCL method with DF observations. However, this novel method only uses observations from high-precision geodetic GNSS receivers for testing, and the quality of SF observations is significantly better than that of low-cost GNSS receivers. To further validate the feasibility and advantages of the proposed method in truly low-cost ionospheric modeling, the future work will focus on the performance evaluation of low-cost GNSS monitoring networks.

Author Contributions: Conceptualization, A.W.; methodology, A.W. and Y.Z.; software, A.W., Y.Z. and J.C.; data curation, A.W., Y.X., J.L. and Y.Y.; validation, Y.X., J.L. and Y.Y.; formal analysis, A.W., Y.Z. and H.W.; writing—original draft preparation, A.W.; writing—review and editing, Y.Z., X.L. and H.W.; supervision, J.C.; funding acquisition, A.W., Y.Z., J.C., H.W. and X.L. All authors have read and agreed to the published version of the manuscript.

Funding: This research was funded by the National Natural Science Foundation of China (42304023; 12403075; 42474034; 42304015; 42274044), the China Postdoctoral Science Foundation (2022M723404), the Fundamental Research Funds for the Central Universities (2024ZKPYDC06), the Program of Shanghai Academic/Technology Research Leader (20XD1404500), the Jiangsu Province Natural Science Foundation of China (BK20231087), the Open Research Foundation of the State Key Laboratory of Geodesy and Earth's Dynamics (SKLGED2024-3-7), and the Innovation Training Program for College Students at China University of Mining and Technology-Beijing (202402002).

Data Availability Statement: The software for the IW SFPPP is the open-source Net_Diff, provided by SHAO via http://202.127.29.4/shao_gnss_ac/ (accessed on 8 February 2025). The final precise satellite orbit and clock products are provided by GFZ via ftp://ftp.gfz-potsdam.de/pub/GNSS/products/mgex/ (accessed on 8 February 2025). The multi-GNSS observations are provided by IGS via ftp://igs.ign.fr/pub/igs/data/campaign/mgex/daily/rinex3/yyyy/ddd/ (accessed on

8 February 2025). The DCB products are provided by CAS via ftp://igs.ign.fr/pub/igs/products/mgex/dcb/ (accessed on 8 February 2025). The GIM products are provided by CODE via https://cddis.nasa.gov/archive/gnss/products/ionosphere/ (accessed on 8 February 2025). The Kp index data are provided by the World Data Center (WDC) via https://wdc.kugi.kyoto-u.ac.jp/wdc/Sec3.html (accessed on 8 February 2025). Information on F10.7 solar radio flux is available from https://wdc.kugi.kyoto-u.ac.jp/kp/index.html#LIST (accessed on 8 February 2025).

Acknowledgments: The authors would like to thank the IGS for the provision of GNSS observation data, precise orbit and clock products, Klobuchar model coefficients, and GIM and DCB products.

Conflicts of Interest: The authors declare no conflicts of interest.

References

1. Feltens, J. The activities of the ionosphere working group of the International GPS Service (IGS). *GPS Solut.* **2003**, *7*, 41–46. [CrossRef]
2. Hernandez-Pajares, M.; Juan, J.M.; Sanz, J.; Aragon-Angel, A.; Garcia-Rigo, A.; Salazar, D.; Escudero, M. The ionosphere: Effects, GPS modeling and the benefits for space geodetic techniques. *J. Geod.* **2011**, *85*, 887–907. [CrossRef]
3. de Bakker, P.F.; Tiberius, C.C.J.M. Real-time multi-GNSS single-frequency precise point positioning. *GPS Solut.* **2017**, *21*, 1791–1803. [CrossRef]
4. Odolinski, R.; Teunissen, P.J.G. An assessment of smartphone and low-cost multi-GNSS single-frequency RTK positioning for low, medium and high ionospheric disturbance periods. *J. Geod.* **2018**, *93*, 701–722. [CrossRef]
5. Wang, A.; Zhang, Y.; Chen, J.; Wang, H.; Yuan, D.; Jiang, J.; Zhang, Z. Investigating the contribution of BDS-3 observations to multi-GNSS single-frequency precise point positioning with different ionospheric models. *Adv. Space Res.* **2024**, *73*, 553–570. [CrossRef]
6. Cahuasquí, J.A.; Hoque, M.M.; Jakowski, N. Positioning performance of the Neustrelitz total electron content model driven by Galileo Az coefficients. *GPS Solut.* **2022**, *26*, 93. [CrossRef]
7. Liu, Z.; Yang, Z. Anomalies in broadcast ionospheric coefficients recorded by GPS receivers over the past two solar cycles (1992–2013). *GPS Solut.* **2016**, *20*, 23–37. [CrossRef]
8. Klobuchar, J.A. Ionospheric time-delay algorithm for single-frequency GPS users. *IEEE Trans. Aerosp. Electron. Syst.* **1987**, *23*, 325–331. [CrossRef]
9. Montenbruck, O.; González Rodríguez, B. NeQuick-G performance assessment for space applications. *GPS Solut.* **2019**, *24*, 13. [CrossRef]
10. Yuan, Y.; Wang, N.; Li, Z.; Huo, X. The BeiDou global broadcast ionospheric delay correction model (BDGIM) and its preliminary performance evaluation results. *Navigation* **2019**, *66*, 55–69. [CrossRef]
11. Wang, A.; Chen, J.; Zhang, Y.; Meng, L.; Wang, B.; Wang, J. Evaluating the impact of CNES real-time ionospheric products on multi-GNSS single-frequency positioning using the IGS real-time service. *Adv. Space Res.* **2020**, *66*, 2516–2527. [CrossRef]
12. Liu, H.; Ren, X.; Xu, G. Investigating the performance of IGS real-time global ionospheric maps under different solar conditions. *Remote Sens.* **2023**, *15*, 4661. [CrossRef]
13. Roma-Dollase, D.; Hernández-Pajares, M.; Krankowski, A.; Kotulak, K.; Ghoddousi-Fard, R.; Yuan, Y.; Li, Z.; Zhang, H.; Shi, C.; Wang, C.; et al. Consistency of seven different GNSS global ionospheric mapping techniques during one solar cycle. *J. Geod.* **2018**, *92*, 691–706. [CrossRef]
14. Rovira-Garcia, A.; Juan, J.; Sanz, J.; Gonzalez-Casado, G.; Ibanez, D. Accuracy of ionospheric models used in GNSS and SBAS: Methodology and analysis. *J. Geod.* **2016**, *90*, 229–240. [CrossRef]
15. Liu, T.; Zhang, B.; Yuan, Y.; Li, M. Real-time precise point positioning (RTPPP) with raw observations and its application in real-time regional ionospheric VTEC modeling. *J. Geod.* **2018**, *92*, 1267–1283. [CrossRef]
16. Huo, X.; Long, Y.; Liu, H.; Yuan, Y.; Liu, Q.; Li, Y.; Liu, M.; Liu, Y.; Sun, W. A novel ionospheric TEC mapping function with azimuth parameters and its application to the Chinese region. *J. Geod.* **2024**, *98*, 13. [CrossRef]
17. Ciraolo, L.; Azpilicueta, F.; Brunini, C.; Meza, A.; Radicella, S. Calibration errors on experimental slant total electron content (TEC) determined with GPS. *J. Geod.* **2007**, *81*, 111–120. [CrossRef]
18. Ren, X.; Chen, J.; Li, X.; Zhang, X. Multi-GNSS contributions to differential code biases determination and regional ionospheric modeling in China. *Adv. Space Res.* **2020**, *65*, 221–234. [CrossRef]
19. Wang, J.; Huang, G.; Zhou, P.; Yang, Y.; Zhang, Q.; Gao, Y. Advantages of uncombined precise point positioning with fixed ambiguity resolution for slant total electron content (STEC) and differential code bias (DCB) estimation. *Remote Sens.* **2020**, *12*, 304. [CrossRef]

20. Zhang, B.; Ou, J.; Yuan, Y.; Li, Z. Extraction of line-of-sight ionospheric observables from GPS data using precise point positioning. *Sci. China Earth Sci.* **2012**, *55*, 1919–1928. [CrossRef]
21. Xiang, Y.; Gao, Y.; Shi, J.; Xu, C. Consistency and analysis of ionospheric observables obtained from three precise point positioning models. *J. Geod.* **2019**, *93*, 1161–1170. [CrossRef]
22. Ren, X.; Chen, J.; Li, X.; Zhang, X. Ionospheric total electron content estimation using GNSS carrier phase observations based on zero-difference integer ambiguity: Methodology and assessment. *IEEE Trans. Geosci. Remote Sens.* **2020**, *59*, 817–830. [CrossRef]
23. Wang, A.; Zhang, Y.; Chen, J.; Liu, X.; Wang, H. Regional real-time between-satellite single-differenced ionospheric model establishing by multi-GNSS single-frequency observations: Performance evaluation and PPP augmentation. *Remote Sens.* **2024**, *16*, 1511. [CrossRef]
24. Su, K.; Jin, S. A novel GNSS single-frequency PPP approach to estimate the ionospheric TEC and satellite pseudorange observable-specific signal bias. *IEEE Trans. Geosci. Remote Sens.* **2022**, *60*, 5801712. [CrossRef]
25. Zhao, C.; Yuan, Y.; Zhang, B.; Li, M. Ionosphere sensing with a low-cost, single-frequency, multi-GNSS receiver. *IEEE Trans. Geosci. Remote Sens.* **2019**, *57*, 881–892. [CrossRef]
26. Li, B.; Zang, N.; Ge, H.; Shen, Y. Single-frequency PPP models: Analytical and numerical comparison. *J. Geod.* **2019**, *93*, 2499–2514. [CrossRef]
27. Wang, N.; Yuan, Y.; Li, Z.; Montenbruck, O.; Tan, B. Determination of differential code biases with multi-GNSS observations. *J. Geod.* **2016**, *90*, 209–228. [CrossRef]
28. Wilgan, K.; Hadas, T.; Hordyniec, P.; Bosy, J. Real-time precise point positioning augmented with high-resolution numerical weather prediction model. *GPS Solut.* **2017**, *21*, 1341–1353. [CrossRef]
29. Chen, J.; Zhang, Y.; Yu, C.; Wang, A.; Song, Z.; Zhou, J. Models and performance of SBAS and PPP of BDS. *Satell. Navig.* **2022**, *3*, 4. [CrossRef]
30. Hernandez-Pajares, M.; Juan, J.M.; Sanz, J.; Orus, R.; Garcia-Rigo, A.; Feltens, J.; Komjathy, A.; Schaer, S.C.; Krankowski, A. The IGS VTEC maps: A reliable source of ionospheric information since 1998. *J. Geod.* **2009**, *83*, 263–275. [CrossRef]
31. Zhou, C.; Yang, L.; Li, B.; Balz, T. M_GIM: A MATLAB-based software for multi-system global and regional ionospheric modeling. *GPS Solut.* **2023**, *27*, 42. [CrossRef]
32. Sanz, J.; Miguel, J.J.; Rovira-Garcia, A.; Gonzalez-Casado, G. GPS differential code biases determination: Methodology and analysis. *GPS Solut.* **2017**, *21*, 1549–1561. [CrossRef]
33. Zhang, Y.; Chen, J.; Gong, X.; Chen, Q. The update of BDS-2 TGD and its impact on positioning. *Adv. Space Res.* **2020**, *65*, 2645–2661. [CrossRef]
34. Wu, J.T.; Wu, S.C.; Hajj, G.A.; Bertiger, W.I.; Lichten, S.M. Effects of antenna orientation on GPS carrier phase. *Manuscr. Geod.* **1993**, *18*, 91–98. [CrossRef]
35. Petit, G.; Luzum, B. *IERS Conventions*; IERS Technical 2010 Note 36; Verlag des Bundesamts für Kartographie und Geodäsie: Frankfurt am Main, Germany, 2010.
36. Boehm, J.; Moller, G.; Schindelegger, M.; Pain, G.; Weber, R. Development of an improved empirical model for slant delays in the troposphere (GPT2w). *GPS Solut.* **2015**, *19*, 433–441. [CrossRef]
37. Wang, A.; Zhang, Y.; Chen, J.; Wang, H. Improving the (re-)convergence of multi-GNSS real-time precise point positioning through regional between-satellite single-differenced ionospheric augmentation. *GPS Solut.* **2022**, *26*, 39. [CrossRef]
38. Lu, Y.; Yang, H.; Li, B.; Li, J.; Xu, A.; Zhang, M. Analysis of characteristics for inter-system bias on multi-GNSS undifferenced and uncombined precise point positioning. *Remote Sens.* **2023**, *15*, 2252. [CrossRef]
39. Feltens, J.; Angling, M.; Jackson-Booth, N.; Jakowski, N.; Hoque, M.; Hernández-Pajares, M.; Aragón-Àngel, A.; Orús, R.; Zandbergen, R. Comparative testing of four ionospheric models driven with GPS measurements. *Radio Sci.* **2011**, *46*, 6. [CrossRef]

remote sensing

MDPI

Article

New Advances Towards Early Warning Systems in the Mediterranean Sea Using the Real-Time RING GNSS Research Infrastructure

Pietro Miele [1,*], Antonio Avallone [1], Luigi Falco [1], Ciriaco D'Ambrosio [1], Shi Du [2], Maorong Ge [3], Roberto Devoti [4], Nicola Angelo Famiglietti [1], Carmine Grasso [1], Grazia Pietrantonio [4], Raffaele Moschillo [1] and Annamaria Vicari [1]

1 Sezione Irpinia, Istituto Nazionale di Geofisica e Vulcanologia, 83035 Grottaminarda, Italy; antonio.avallone@ingv.it (A.A.); luigi.falco@ingv.it (L.F.); ciriaco.dambrosio@ingv.it (C.D.); nicola.famiglietti@ingv.it (N.A.F.); carmine.grasso@ingv.it (C.G.); raffaele.moschillo@ingv.it (R.M.); annamaria.vicari@ingv.it (A.V.)

2 Deutsches GeoForschungsZentrum, 14473 Potsdam, Germany; shidu@gfz-potsdam.de

3 Shanghai Astronomical Observatory, Shanghai 200030, China; maorongge@shao.ca.cn

4 Osservatorio Nazionale Terremoti, Istituto Nazionale di Geofisica e Vulcanologia, 00143 Rome, Italy; roberto.devoti@ingv.it (R.D.); grazia.pietrantonio@ingv.it (G.P.)

* Correspondence: pietro.miele@ingv.it

Highlights:

What are the main findings?

- For the first time an assessment of the accuracies of the RING real-time GNSS position time series is performed, by comparing GPS and GNSS solutions and different Precise Point Positioning (PPP) strategies (standard PPP and PPP with Regional Augmentation PPP-RA)
- The most accurate (sub-centimeter level) RING real-time solutions are obtained for GNSS solutions with the PPP-RA strategy by using a 300 s sliding window for solutions.

What are the implications of the main findings?

- The real-time RING GNSS infrastructure can potentially observe co-seismic ground deformation after moderate-magnitude (M6) earthquakes.
- The proposed approach supports the development of effective tsunami and earthquake early warning systems in tectonically active regions.

Abstract: Nowadays, information obtained through Global Navigation Satellite Systems (GNSSs) is widely employed in modern geodesy. The Precise Point Positioning (PPP) approach, which leverages signals from multiple GNSS constellations (e.g., GPS, GLONASS, Galileo, and BeiDou), enables high-precision positioning—crucial for seismic monitoring and early tsunami warning systems (EEWs). Recent advances, such as increased satellite availability and additional frequency bands, have significantly improved PPP performance, particularly in terms of positioning accuracy and convergence time. This study focuses on the Rete Integrata Nazionale GNSS (RING) network, managed by the Istituto Nazionale di Geofisica e Vulcanologia (INGV), which comprises dual-frequency GNSS receivers distributed across the Italian peninsula and parts of the Mediterranean Basin. We evaluate the performance of the RING data (GPS and GNSS) acquired in a period of three weeks between 19 January 2024 and 9 February 2024 and analyzed in real time by using different PPP strategies: standard PPP and PPP with Regional Augmentation (PPP-RA). The preliminary results show that the PPP-RA approach enhances positioning accuracy and reduces convergence time, especially when comparing GPS-only datasets with those incorporating full multi-GNSS configurations. For the *daily* solution, in the optimal setup (i.e., full GNSS with RA), real-time solutions exhibit average accuracies of 2.05, 1.73, and 4.35 cm for the North, East, and vertical components, respectively. *Sub-daily* accuracies' analysis, using 300 s sliding windows, showed even better uncertainties, exhibiting median values of 0.41, 0.32, and 0.9 cm for the North, East and vertical components, respectively. Based on the outcomes for network-wide sub-daily accuracies, 84% of the stations demonstrate average errors within 2 cm for North and East components and 3 cm for the vertical one. The analysis on the convergence time after data gaps occurred during the investigation period shows that 87% of the RING stations experienced convergence times lower than five minutes in the GNSS PPP-RA solution. These findings underscore the potential of RT-GNSS RING data for enhancing seismic monitoring and early warning systems, particularly in tectonically active regions.

Keywords: GPS; GNSS; PPP-RA; RING network

1. Introduction

Modern geodesy utilizes a range of space and terrestrial technologies that contribute to the investigation and understanding of the solid earth, atmosphere, and oceans [1]. Space geodesy, among many functions, provides observations that form the basis of physical models of the earthquake deformation cycle, offering key information to describe the processes leading up to and following significant events based on deformation rates on the Earth's surface. During an earthquake, the antenna, anchored to the ground, moves, and these position variations can be calculated in real time (RT) to monitor both dynamic and static displacement. The Global Navigation Satellite System (GNSS) displacement waveforms provide reliable centimeter-level accurate representations of the ground motion for moderate-to-large (M > 6) seismic events and at local-to-regional scales, e.g., [2–5]. For these events, as displacements represent native measurements for GNSSs, when available, they should be considered preferable with respect to doubly integrated accelerograms due to well-known baseline offset problems [6]. The GNSS technique has been increasingly applied in seismic monitoring using the Precise Point Positioning (PPP) approach [7], enabling the direct computation of co-seismic displacements, rapid and accurate magnitude determination [8,9], and moment tensor calculation [10,11]. Consequently, in recent years, RT-GNSS data has been integrated into both earthquake and tsunami early warning systems (EEW-TEW), particularly for large events [12–16].

The PPP technique is based on the measurements of a single receiver observing a number of satellites and capable of obtaining high-precision positions. However, it requires a convergence time of tens of minutes up to several hours [17,18]. Another factor that affects the time to reach centimeter-level accuracy is the number and geometry of visible satellites for a given portion of the Earth's surface. However, in recent years, PPP has been able to take advantage of more frequencies and satellite resources due to the restoration of the complete orbital constellation for GLONASS and the emergence of new satellite systems such as Galileo and BeiDou (BDS). The combination of different GNSS constellations increases the number of visible satellites and improves their geometry, presenting the possibility of improving the performance of the PPP in terms of the positioning accuracy and convergence time [19–21].

Thanks to these achievements, RT-GNSS data have also been employed in early tsunami warning systems [22–24] using two real-time observational networks (GNSS and seismic), leading to a denser global coverage for earthquake finite-source inversion. However, to achieve this, high-rate GNSS data and a solid processing infrastructure are essential. Over the past three decades, numerous permanent GNSS networks have been established worldwide [25–27]. Since 2004, the Istituto Nazionale di Geofisica e Vulcanologia (INGV) has operated a continuous network to observe and monitor the active tectonic processes in Italy [28,29]. The Italian peninsula, due to its tectonic evolution driven by the interaction between the African and Eurasian plates, represents a unique natural laboratory for geodynamical investigations. The region is characterized by complex tectonic and geological settings, with several seismically active domains, particularly in the Apennines and the northeastern Alps [30–32].

At the beginning, the abovementioned network, known as the Rete Integrata Nazionale GPS (RING), was equipped with instrumentation that could only acquire data from the Global Positioning System (GPS) constellation. Nowadays, this important research infrastructure is mostly composed of dual-frequency GNSS (GPS, GLONASS, GALILEO, and BEIDOU) receivers and antennas. The RING network currently comprises approximately 250 stations, mostly distributed across Italy, with additional sites in Malta and Greece.

At present, the use of real-time PPP in high-precision seismic monitoring remains challenging due to the relatively lower quality of real-time satellite orbits and clocks compared to precise final (post-processed) products [33]. From a technical point of view, a real-time GNSS product chain is typically structured with a server/client configuration, and it can be vulnerable to connection interruptions. Moreover, high-precision GNSS positioning requires correction streams (i.e., satellite orbits and clocks) computed by the external agency servers based on remote or global reference networks.

To overcome these limitations, scientifically advanced real-time PPP approaches should be considered. In addition to the standard PPP method, these strategies also include a single-receiver ambiguity resolution by applying corrections derived from a local reference network (i.e., the PPP-RA [34,35]) or from a global network (i.e., the RTPPP PPP-AR [36,37], PPP-Wizard [18]) and single-receiver Regional Augmentation (i.e., the RTPPP PPP-RA [38]). In fact, both ambiguity resolutions and Regional Augmentations are crucial for reaching high-accuracy products and fast convergence times after the occurrence of data gaps [39]. Moreover, multi-GNSS PPP outperforms the single system in both convergence and precision, as demonstrated by ambiguity-float and also fixed PPP [40].

In this study, following an initial overview of the implementation and management of the RING network, the performance of the GNSS infrastructure is assessed based on preliminary results from PPP-RA processing. Specifically, its measurement accuracy is evaluated through the analysis of the real-time datasets. In this study, a GPS-only dataset is compared with a full GNSS configuration in a research software environment, namely RTPPP (v1.4).

2. Materials and Methods

2.1. Real-Time RING Data Acquisition

Since 2006, the INGV has run the RING network (https://ring.gm.ingv.it, (accessed on 30 June 2025)) with the aim of studying deformations at different spatial and temporal scales in the Eurasia–Africa plate boundary. With regard to the different spatial scales, the purpose is to observe and model how the deformation is localized or distributed from a single seismogenic fault (or fault systems) to regional or plate-scale areas. With regard to studying deformation at different temporal scales, the purpose is to explore the whole frequency spectrum of the deformation up to sampling frequencies useful for earthquake source studies [40–43]. In this sense, the availability of real-time time series at RING sites would significantly contribute to seismological early warning applications. Nowadays, the whole RING network is composed of 253 stations (Figure 1, white triangles), whose spatial distribution is not homogeneous all over Italy, but instead it is denser where the accumulation of the deformation is expected, i.e., where the main seismogenic faults are present. Among these 253 stations, currently, about 150 sites are set up to transmit RT-GNSS data towards the RING acquisition center located at the INGV Irpinia headquarters (Grottaminarda, Southern Italy).

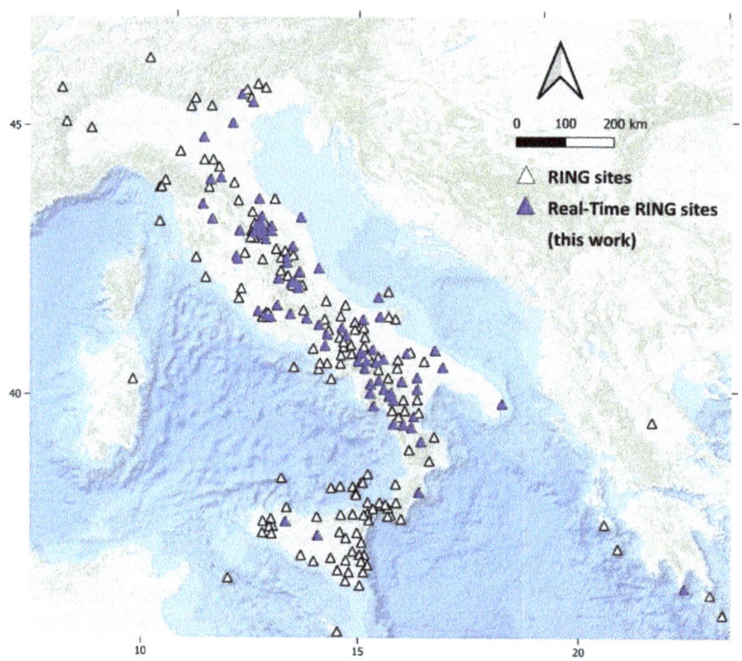

Figure 1. The spatial distribution of the RING network: each marker corresponds to a GPS/GNSS station. The blue filled triangles represent sites equipped with GNSS receivers considered in this study.

The RING data is primarily transmitted through LTE [44] and Wi-Fi technologies. Additionally, a significant number of RING sites also transmit data via satellite telemetry, as they are co-located with broadband and accelerometer instrumentation from the Rete Sismica Nazionale (RSN https://terremoti.ingv.it/instruments/network (accessed on 10 July 2025)), another INGV research infrastructure on which the seismic monitoring of Italy and surrounding countries is based [45]. Real-time RING data streams, formatted according to the standard RTCM v.3, are acquired and managed via a specifically tuned Ntrip Caster, developed by BKG (https://igs.bkg.bund.de/ntrip/bkgcaster (accessed on 10 July 2025)), hereafter RING Ntrip Caster). Metadata, accessible through the RING website, are also available as a source table, regularly synchronized with the RING database.

2.2. Real-Time Data Analysis

All the RING GNSS data as well as those belonging to other permanent networks operating in the Mediterranean region are routinely processed using various scientific software packages, such as Bernese (v5.2), GAMIT/GLOBK (v10.7), and GIPSY-OASIS (v6.3), to observe long-term deformation in the Eurasia–Africa plate boundary region (https://webring.gm.ingv.it/index.php/data-analysis-and-product/ (accessed on 25 July 2025)) [46]. The processing approach adopted in the first two software packages (Bernese and GAMIT/GLOBK) is based on the relative positioning concept using phase-observable double-differencing techniques, whereas the third processing scheme (GIPSY-OASIS) is based on the Precise Point Positioning (PPP) technique [7,47,48].

The RING real-time GNSS data analysis presented in this study was carried out using the RTPPP software (v1.4), developed by the German Research Centre for Geosciences (GFZ) [39]. The RTPPP is a high-precision system for monitoring the real-time displacement of reference stations. It uses undifferenced, dual-frequency code and carrier-phase observations with available real-time satellite orbits and clock products to achieve positioning results with a centimeter-level precision at each station [39]. The RTPPP environment consists of three parts: (i) PPP client software, (ii) a database, and (iii) a website (Figure 2) (https://www.gfz.de/en/section/space-geodetic-techniques/projects/rt-ppp-real-time-gnss-precise-positioning (accessed on 10 June 2025)). In this work we will focus on the first part (i) to assess the accuracy of the real-time RTPPP products for the RING network. The PPP client supports different data sources, including real-time replay and post-processing. Data processing can be conducted considering multi-GNSS constellations, including the GPS, GLONASS, Galileo, and BDS. Different positioning modes are supported, e.g., fixed coordinate, static, and kinematic modes [49].

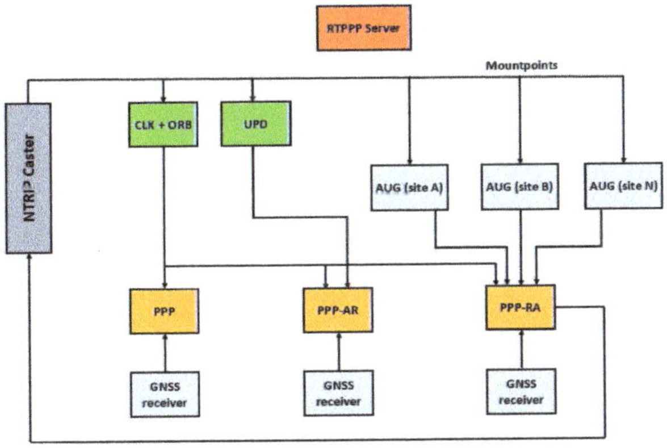

Figure 2. Overall simplified structure of RTPPP software (modified from [49]).

Following the strategies proposed in previous works dealing with the development of the RTPPP software [38,50], the real-time RING data analysis is briefly described, as the main focus of this paper is related to the accuracy assessment of the RING real-time time series. We followed three main steps (Figure 2). In the first step, for each RING station, PPP-float position time series (defined PPP solution) were obtained by analyzing the raw data streaming by using external real-time satellite products (orbits and clocks) available on the IGS portal (https://igs.org/rts/products/ (accessed on 10 June 2025)), the ionosphere-free linear combination was used to reduce the first-order ionosphere effect, the GMF mapping function was used for troposphere ZTD estimations, and the IGS14 absolute antenna calibration was used for better modeling of the antennas delays. In this first step, the phase integer ambiguities were estimated and uploaded, in real time (UPD), to the RING NTRP Caster (Section 2.1). In the second step, these UDP streams were used, in addition to the real-time satellite orbits and clock products and the raw data streaming from the stations, to resolve ambiguities station by station. By applying the UPD corrections, the atmospheric correction (AUG) of the ionospheric slant and troposphere zenith wet delay were derived for each station from the PPP-fixed solutions and uploaded to the RING Ntrip Caster. This includes the use of a real-time spatial common mode for each station, estimated by using data from at least three other closer and surrounding sites (defined PPP-RA solution). Further details on the UPD and AUG estimations can be found in previous works [39,49].

By applying UPD corrections, integer undifferenced ambiguities on receiver frequencies can be fixed in PPP mode at all regional monitoring stations. The constraints imposed on the kinematic coordinates of adjacent epochs are fine-tuned using an adaptive filter in real time to strengthen the solution. Moreover, the atmospheric delay is often quite stable over short periods and can be represented by a constant or linear function. Therefore, even in periods of strong shaking, the station position and atmosphere are distinguishable during parameter estimation because of the significant difference in their temporal characters. For regional reference networks with moderate-to-short baselines (an inter-station distance of a few tens of kilometers), a cm-level accuracy can be achieved for the interpolated atmospheric delay corrections. These corrections are imposed as a strong constraint on the related parameters of the monitoring station, while the coordinates are estimated in kinematic mode. By applying this atmospheric delay model (AUG) to the PPP-fixed solution, in addition to the orbit and clock products used, both an instantaneous ambiguity resolution and Regional Augmentation can be achieved at the monitoring station [38,50]. In the RTPPP environment, this latest PPP solution is named PPP-RA.

The aim of this work is to compare GPS-only and GNSS real-time solutions developed from RING data by using different PPP strategies: standard PPP and PPP with Regional Augmentation (PPP-RA). Our investigation period is from 19 January to 9 February 2024 (three weeks). For all the solutions, we used products derived from multi-constellation data available through the IGS portal (https://igs.org/rts/ (accessed on 10 June 2025)). Specifically, we used the CLK93 (APC) stream for SSR orbit/clock corrections and the RTCM3EPH-MGEX stream for multi-GNSS broadcast ephemeris.

2.3. Data Accuracy Estimation

To assess the positioning accuracy of real-time estimated PPP time series, we adopted two different strategies:

- In the first approach, for each station, RMS values for the East, North, and Up components were computed using all epochs acquired throughout the day. The RMS values were calculated considering all the epochs for which the position had been estimated (up to 86,400 at a 1 Hz sampling frequency for a 24 h session). In addition, we also distinguished the RMS calculated for the epochs in which Regional Augmentation (RA) had been applied to the positioning. These two different approaches were carried out for both GPS-only and full GNSS solutions, resulting in a single RMS value per component, per station, and per day of measurement. These RMS values will hereafter be referred to as *"daily RMS"*.
- In the second approach, RMS values were computed for each time series (i.e., each component) using sub-daily overlapping sliding windows. While this method generally yields lower RMS values due to the smaller number of samples, it is particularly useful for detecting sudden transient deformations, such as co-seismic static and dynamic displacements, over shorter time scales. Given the time duration of the co-seismic shaking (30–60 s) during most of the past moderate-magnitude earthquakes in Italy (references), we decided to use 300 s sliding windows with an overlap of 150 s for each step. The 300 s time window is considered enough to sample either the noise or co-seismic dynamic and static displacements for moderate-magnitude earthquakes (M6). The overlaps are used to increase the number of RMS estimations. These RMS values will hereafter be named *"sub-daily RMS"*.

Both strategies were applied to assess the accuracy of the following datasets: (a) standard PPP GPS solutions; (b) PPP-RA GPS solutions; (c) standard PPP GNSS solutions; and (d) PPP-RA GNSS solutions.

Moreover, the performance of the RING network was evaluated using the average RMS, allowing us to categorize GNSS sites in terms of performance and reliability over the time span considered.

3. Results

This section is divided by subheadings illustrating the experimental results obtained by following the two previously described approaches.

Taking into account that the convergence time is a key metric for real-time GNSS applications [48], this aspect was investigated for the analyzed datasets. Acquisition gaps may occur due to various factors, such as environmental conditions, technical issues, or interruptions in data transmission. Previous works define [48] the "convergence time" as the time required to reach a positioning error below 10 cm after a gap. In this work, for each component, for each station, and for each solution (GPS PPP, GPS PPP-RA, GNSS PPP, and GNSS PPP-RA) we estimated the defined convergence time after any gap occurring during the whole period of investigation (Section 2.2). The results of this huge analysis are described and shown in the Supplementary Materials. In Figure 3, we show an example for the CONI station.

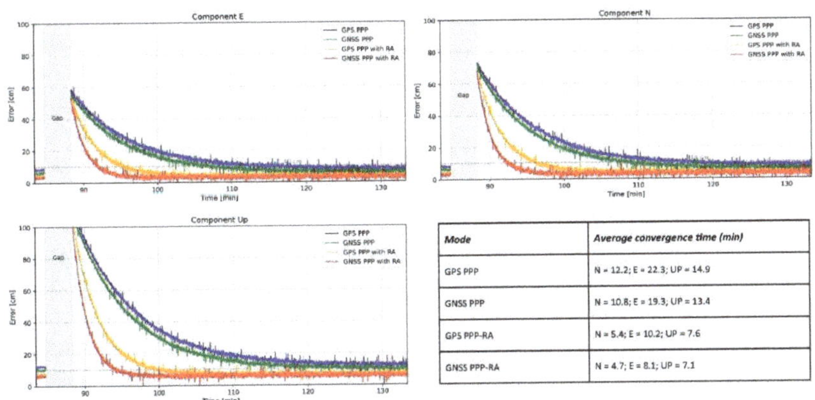

Mode	*Average convergence time (min)*
GPS PPP	N = 12.2; E = 22.3; UP = 14.9
GNSS PPP	N = 10.8; E = 19.3; UP = 13.4
GPS PPP-RA	N = 5.4; E = 10.2; UP = 7.6
GNSS PPP-RA	N = 4.7; E = 8.1; UP = 7.1

Figure 3. Convergence curves of the three positioning components for the CONI station after a gap that occurred on 24 January 2024. The values in the table refer to the average convergence times for the different acquisition modes considered in this study across all RING sites.

The convergence times resumed in Figure 3 may also depend on the characteristics of each acquisition site. Indeed, among the considered RING stations, there is significant variability in the geographical location, which directly affects the contribution of the Regional Augmentation. In Figure 3, we also show a table that displays the average convergence time estimations carried out for all the solutions and for all the RING stations. Most sites (87.5%) show convergence times shorter than five minutes (e.g., CONI site), even when considering GPS-only PPP data (see Supplementary Materials). In any case, it is important to highlight the significant impact of both the larger number of satellites used in the full GNSS solutions and the contribution of the Regional Augmentation for both the GPS-only and full GNSS datasets.

3.1. Daily RMS

The RMS values obtained with the "*Daily RMS* approach" are illustrated in Figure 4, which consists of four subplots showing an example of the daily RMS values computed for all the analyzed RING sites (see Figure 1) and for the following different datasets: Figure 4a, the GPS time series for all the available epochs sampled at 1 Hz, including both PPP and PPP-RA solutions; Figure 4b, the GPS time series sampled at 1 Hz considering only higher-accuracy PPP-RA solutions; Figure 4c, the GNSS the time series sampled at 1 Hz for all the available epochs, including both PPP and PPP-RA solutions; and Figure 4d, the GNSS time series sampled at 1 Hz considering only higher-accuracy PPP-RA solutions. Below each subplot, we have included a graph quantifying the completeness of the considered time series. The graphs in Figure 4a,c show the percentage of the epochs with any available solutions with respect to the expected ones (i.e., 86,400 for a 24 h solution); meanwhile, graphs in Figure 4b,d show the percentage of the epochs with available PPP-RA solutions with respect to the expected (i.e., total) epochs. All the solutions in Figure 4 are for 24 January 2024, and blue, red, and green markers correspond to values for the North, East, and Up components, respectively. As expected, the vertical component is systematically (two to three times) noisier than the horizontal ones, as the Up component is often affected by tropospheric and ionospheric residuals during the analysis. The GPS solution (Figure 4a) shows average accuracies of 2.22, 2.70, and 6.69 cm for the North, East, and Up components, respectively. By applying the Regional Augmentation (PPP-RA) in the GPS solution (Figure 4b), we were able to largely improve the average accuracies: 36% for North (1.42 cm), 54% for East (1.23 cm), and 18% for Up (5.46 cm). A further significant improvement is clearly visible in the full GNSS solutions (Figure 4c,d). Considering all the presented solutions, it is important to underline that there is no clear correlation between *daily RMS* values and their completeness.

The full GNSS solution, including the standard PPP strategy (Figure 4c), exhibited accuracies of 1.58, 1.64, and 3.74 cm for North, East, and Up components, respectively, whereas the full GNSS PPP-RA solution (Figure 4d) exhibited uncertainties of 1.32 cm, 0.97 cm, and 2.97 cm for North, East, and Up, respectively. Comparable accuracies were generally observed for the same types of solutions in the whole dataset. In Figure 4, the time series of the daily RMS values are shown for some examples of sites, spanning all measurement days. The sites are TEOL, PIOB, and AV04, whose locations are shown in the map in Figure 5.

For these sites, the RMS values calculated considering all acquired epochs are significantly higher than those obtained when RA epochs are estimated, for both the GPS-only and GNSS datasets. Improvements are especially appreciable in the vertical component. The advantages of RA products are more evident in the time series for each station, shown in Figure 5, where the dotted lines representing RMS values for RA-only epochs are consistently lower than the solid lines (i.e., RMSs for all epochs). For the horizontal components, the dotted lines are more stable, with values below 2 cm in both the GPS-only and GNSS configurations, while for the vertical component, the enhancement due to RA is substantial, leading to RMS values below 4 cm, particularly with GNSS products.

The frequency distribution of the North (N), East (E), and Up (UP) components over the entire analyzed period is shown in Figure 6.

As can be seen, compared with the median RMS of the total epochs for all three components, the PPP-RA products have lower median RMS values for the N, E, and UP components than the PPP ones. The histograms also depict decreasing average values for all three components. For the GPS-only dataset, the RMS is generally higher, and the difference in median and average values between the histogram that includes all epochs and the one that considers only those with RA is more marked.

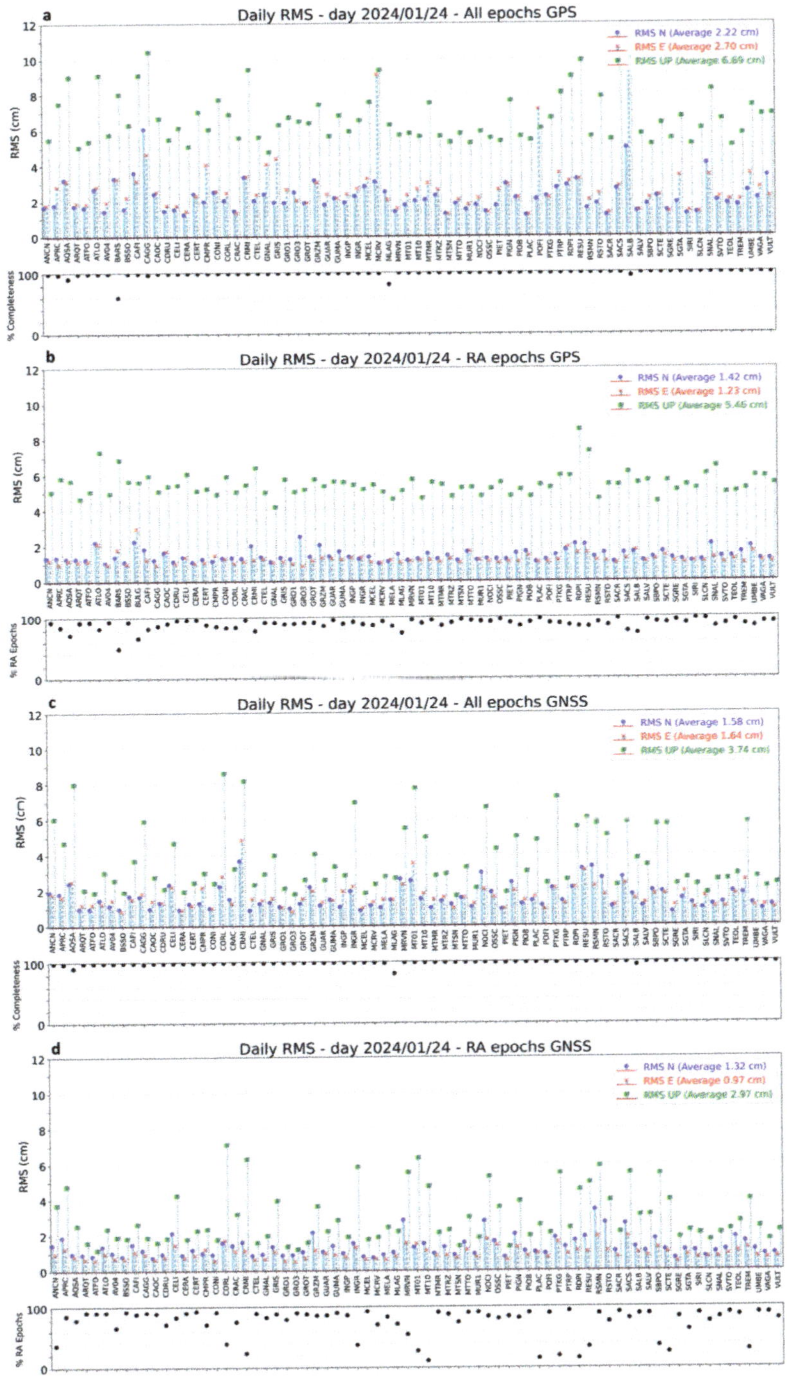

Figure 4. RMS distribution for RING stations on 24 January 2024, considering GPS total epochs (**a**), GPS RA epochs (**b**), GNSS total epochs (**c**), and GNSS RA epochs (**d**). The percentage of acquired epochs and those with Regional Augmentation are shown in the scatter plot below each stem plot.

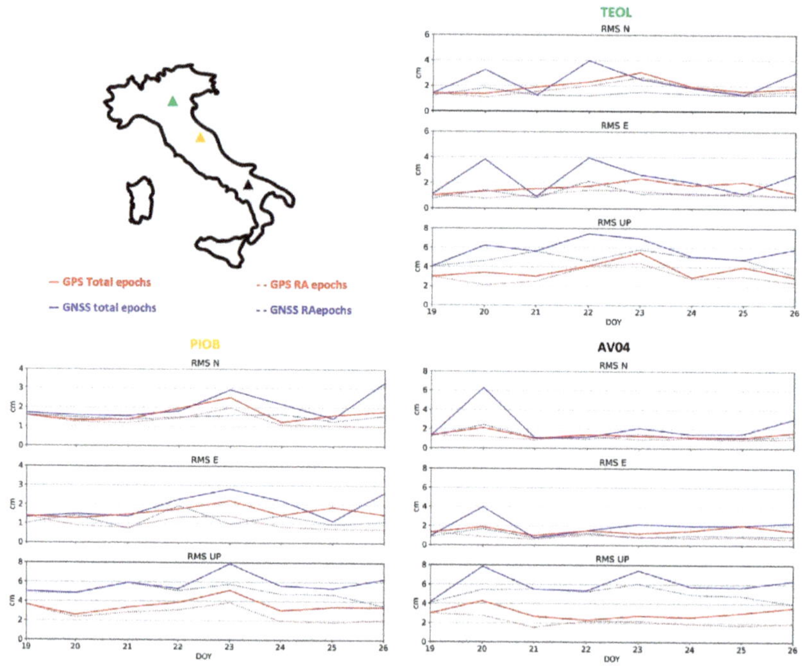

Figure 5. Time series for the TEOL, PIOB, and AV04 stations. The RMS values for the three components were calculated using the daily solutions for the entire period considered. Colored triangles show the locations of the sample stations selected.

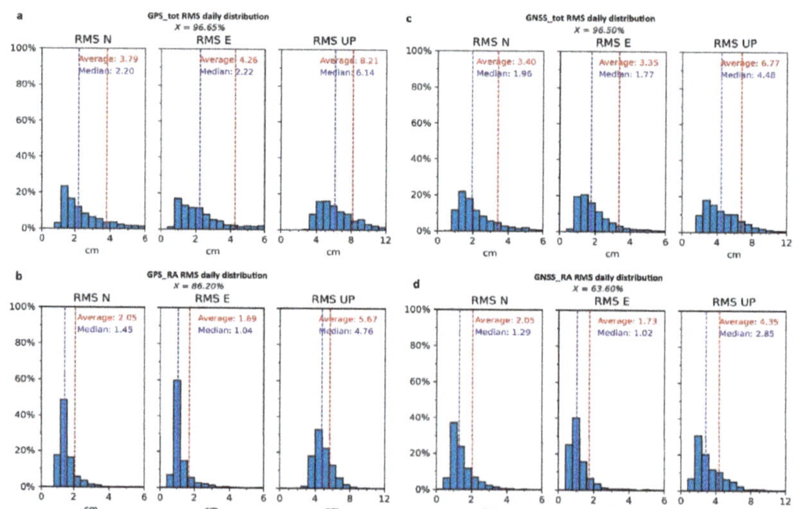

Figure 6. Distribution of daily RMSs over the GPS-only and full GNSS datasets. In the upper part (**a**,**c**), the histograms show the values for all acquired epochs, while in the lower part (**b**,**d**), the histograms represent only the RA epochs. The blue dotted line indicates the median values, and the red line represents the average values. The "X" value in subplots (**a**,**c**) denotes the percentage of acquired epochs during the considered period (3 weeks), while in subplots (**b**,**d**), it represents the percentage of epochs with Regional Augmentation.

3.2. Sub-Daily RMS

We calculated RMS values by splitting the full 24 h daily observations into 300 s sliding windows with a 150 shift period, resulting in 576 sub-daily samples for each acquisition day. The differences between the GPS-only and GNSS datasets were also analyzed.

Figure 7 shows the data for all stations of the network on 24 January 2024, as in Figure 4. As can be clearly seen, the short-term analysis using sliding windows yields significantly smaller RMS values than those calculated for the entire day, even when all epochs are considered (Figure 7a,c).

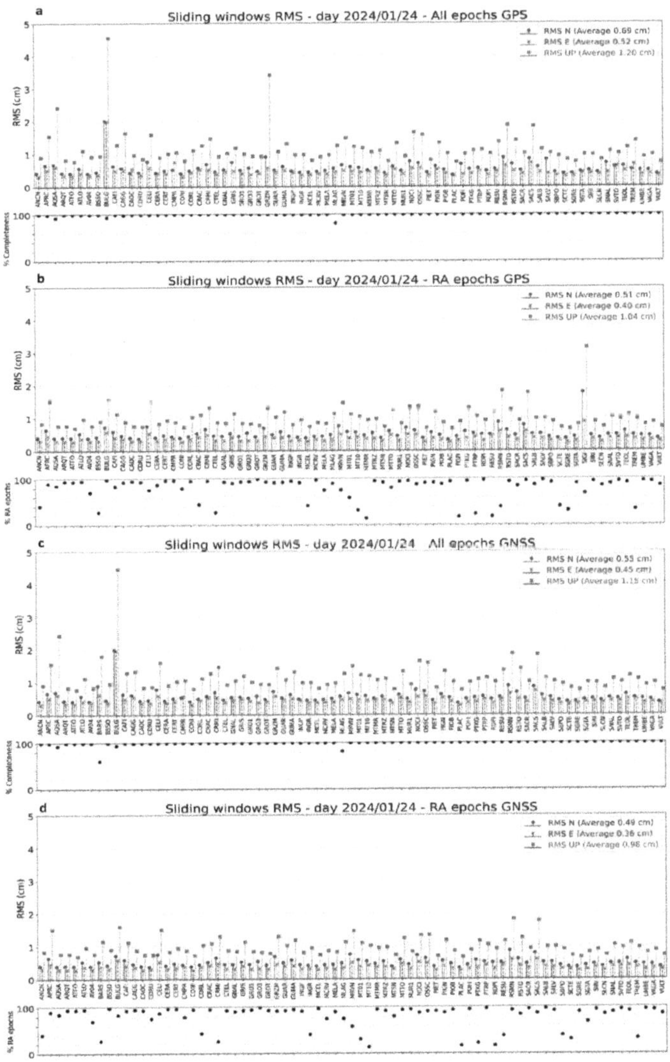

Figure 7. The RMS distribution along the RING network for 24 January 2024, using the sliding windows approach. RMS values computed for the (**a**) PPP GPS solution; (**b**) PPP-RA GPS solution; (**c**) PPP GNSS solution; and (**d**) PPP-RA GNSS solution. The percentage of acquired epochs and those with Regional Augmentation is shown in the scatter plot below each stem plot.

Figure 8 shows the average RMS values computed for the AV04, TEOL, and PIOB stations using the aforementioned sliding windows. Notably, the sliding windows based on RA-only epochs (blue and red dotted lines) demonstrate sub-centimeter accuracy, even for the vertical component in the GNSS RA datasets. However, the GPS-only products also exhibit sub-centimeter RMS values, particularly for the horizontal components.

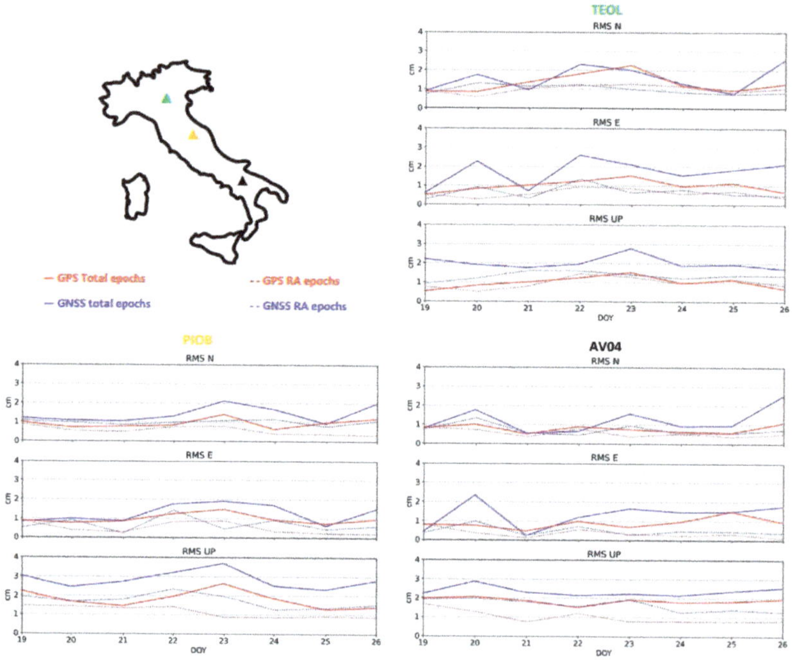

Figure 8. Time series of average RMS value computed on 300 s sliding windows for TEOL, PIOB, and AV04 stations. Colored triangles show the locations of the sample stations selected.

The histograms of the RMS for the three components of the GPS-only dataset are shown in Figure 9a,b, considering all epochs and only those with RA, respectively. All distributions are characterized by sub-centimeter average and median values for the horizontal components and values close to 1 cm for the vertical component.

In the distribution shown in Figure 9a, RMS values are concentrated in the 0–2 cm interval, with a median value of 0.50 cm and 0.41 cm for the North and East components, respectively. For the vertical component, the histogram shows a dense distribution of values within the 0–4 cm range, with a median value of 1.06 cm. When considering only the GPS epochs with RA (Figure 9b), a general leftward shift is observed, with median values of 0.46 cm for the North component, 0.39 cm for the East component, and 1.08 cm for the vertical component.

The distribution of RMS values for the GNSS dataset is shown in Figure 9c,d. As expected, these values calculated within the selected sliding windows using full GNSS solutions are lower than those for the GPS-only dataset. The most favorable results are shown in Figure 9d, which corresponds to the GNSS solution with the Regional Augmentation. The median values are 0.41 cm for the North component, 0.32 cm for the East component, and 0.82 cm for the vertical component.

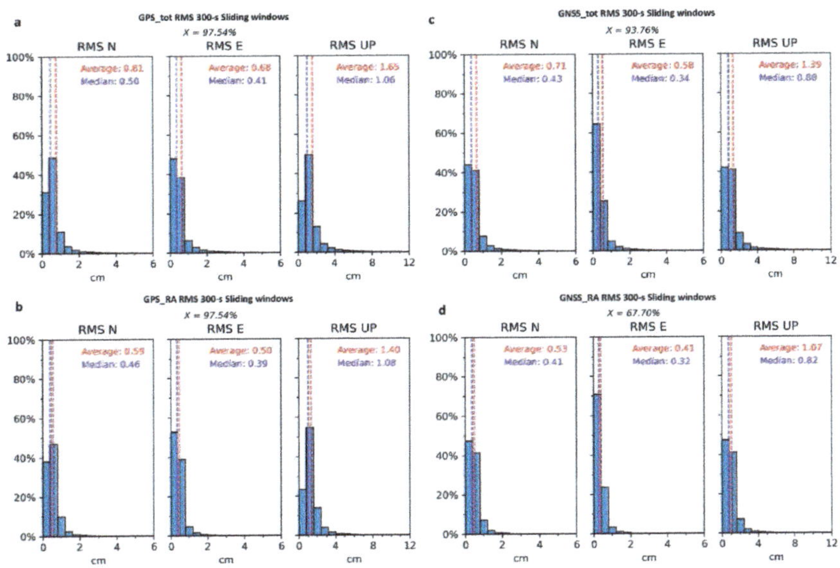

Figure 9. The distribution of the RMS for short-term solutions over the GPS-only and full GNSS datasets: (**a**) GPS total epochs dataset; (**b**) GPS RA epochs dataset; (**c**) GNSS total epochs dataset; and (**d**) GNSS RA epochs dataset. The blue dotted line identifies the median value, and the red dotted line represents the average values. The "X" value for subplots (**a**,**c**) represents the percentage of acquired epochs during the considered period (3 weeks), while for subplots (**b**,**d**), it indicates the percentage of epochs with Regional Augmentation.

3.3. RING Network Overview

Finally, to evaluate the performance of the RING network, it is crucial to identify the stations that demonstrate optimal performance and reliability. In this study, the average daily RMS was selected from different parameters for this purpose. Classification was based on daily values for both the GPS-only and full GNSS datasets, as illustrated by the statistics shown in Figure 4, for example. Furthermore, we compared both GPS-only and full GNSS solutions via a short-term accuracy analysis (sliding windows) considering all the epochs acquired.

The maps in Figure 10 depict the spatial distribution of the analyzed RING stations for both the GPS-only and full GNSS datasets, classified according to the average RMS values for each component.

For the horizontal components (Figure 10a,b,d,e), most RING stations are represented by green and yellow markers, indicating an average RMS of less than 5 cm. However, some stations display significantly higher RMS values (>5 cm), particularly in the GPS-only dataset (Figure 10a,b, orange and red triangles). In contrast, the average vertical RMSs (Figure 10c,f) reveal a different pattern. Specifically, in the GPS-only dataset (Figure 10c), the average value of most stations exceeds 5 cm, as also indicated in the histograms (see Figure 6a). Conversely, in the map in Figure 10f, corresponding to the full GNSS dataset, an overall improvement is observed, with many stations exhibiting average RMSs of just a few centimeters, with the RMSs exceeding 7 cm at only nine stations.

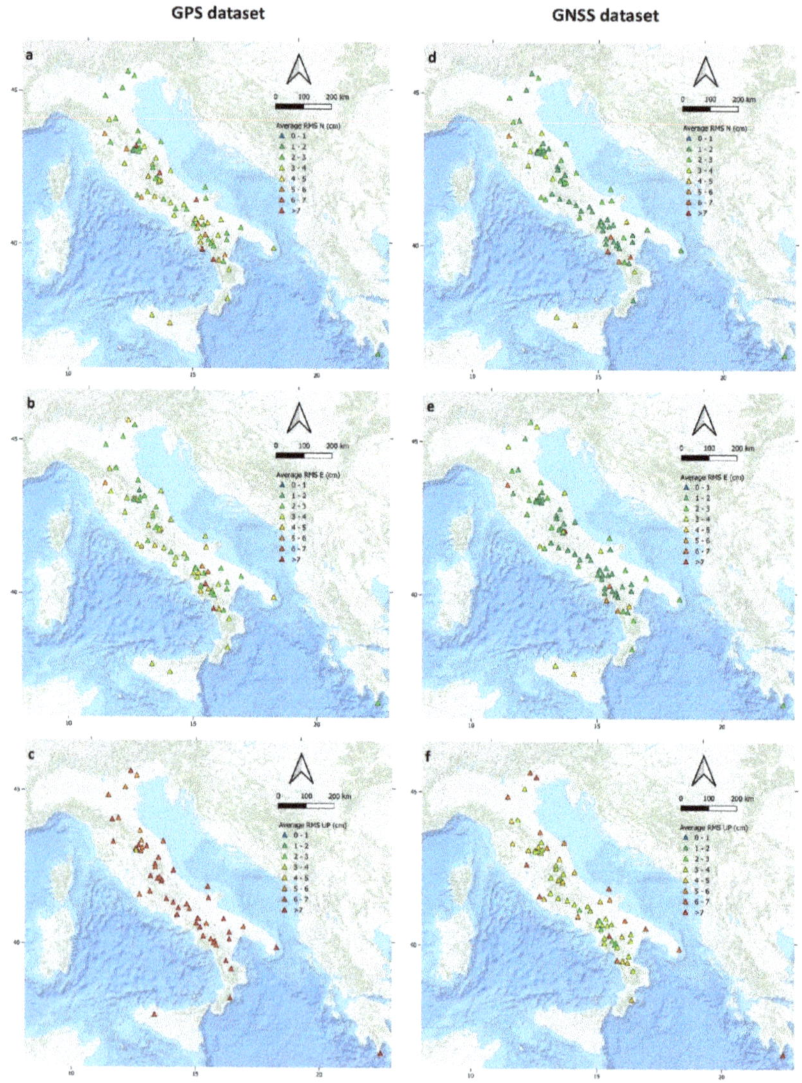

Figure 10. A comparison of the performance of RING stations, taking into account the GPS-only dataset (left column; (**a**–**c**) for N, E, and Up components, respectively) and the full GNSS dataset (right column; (**d**–**f**) for N, E, and Up components, respectively). The average values represented for each site were calculated using the daily solutions covering the considered analysis period.

The comparison between the two datasets clearly underscores the improvements achieved with a full GNSS configuration. This is further illustrated in Figure 11, which shows the spatial distribution of the average RMS, calculated using the sliding windows approach. The maps highlight significantly different scenarios compared to the daily solution analysis. For both the GPS-only and GNSS datasets, nearly all average RMS values fall within the 0–2 cm range (blue and dark green triangles); the accuracies for the vertical component (Figure 11c,f) are also within this range. Overall, this approach demonstrates the potential to achieve a sub-centimeter RMS resolution.

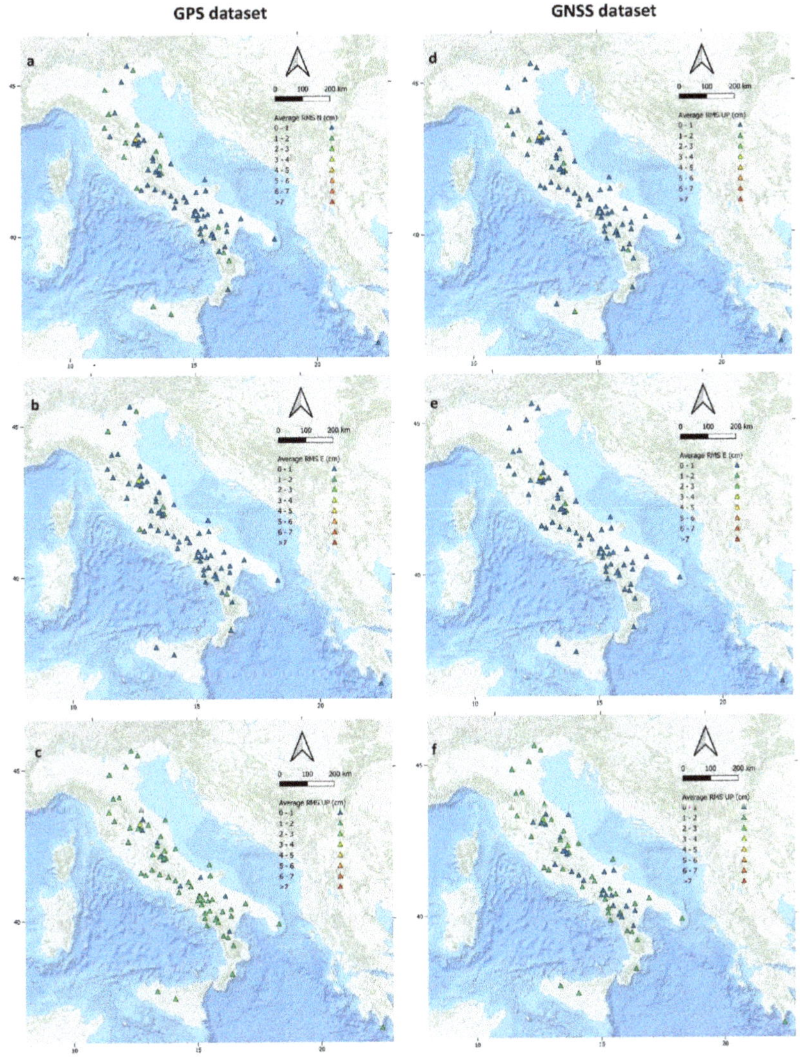

Figure 11. A comparison between the performance of RING stations, taking into account the GPS-only dataset (left column; (**a–c**) for N, E, and Up components, respectively) and the full GNSS dataset (right column; (**d–f**) for N, E, and Up components, respectively), with average RMS values computed using the sliding window approach.

4. Discussion

This paper presents the first assessment of real-time position time series from the RING network. Positioning was performed using the PPP method, which, unlike network-based solutions, relies solely on data from a single designated station and the precise positions of satellites. In this study, a real-time data analysis was conducted with the RTPPP software at 88 RING stations using two distinct datasets: one based on the GPS only and the other based on full GNSS solutions.

For both daily solution datasets, the application of Regional Augmentation proved effective in enhancing the positioning precision for the North, East, and vertical components, achieving values close to the centimeter level. Overall, the comparison between RMSs of

GPS-only and full GNSS datasets showed a general improvement for both horizontal and vertical components employing the latter configuration.

Additionally, we compared GPS-only and full GNSS solutions by performing a short-term accuracy analysis over the entire study period, considering the fact that the main earthquakes in Italy over the past 15 years (Mw 6.3 L'Aquila, 2009; Mw 5.6 and 5.8 Emilia, 2012; Mw 6.0 Amatrice, 2016; and Visso and Norcia, 2016) exhibited dynamic displacements concentrated within just a few tens of seconds [32,50]. Accordingly, accuracy assessments for short time windows are valuable for applications where deformations are concentrated in a very limited temporal window (e.g., detecting co-seismic displacements).

This analysis revealed that sub-centimeter RMS errors can be achieved within short sub-daily windows. The 300 s time windows for full GNSS solutions exhibited median values of 0.41, 0.32, and 0.9 cm for the North, East, and vertical components, respectively. However, the outcomes highlight that a continuous RA contribution is required to ensure stable performance over time; interruptions in the augmentation contribution may lead to degraded and inconsistent solutions (Figure 12). Furthermore, as shown in previous studies, various factors can affect the quality of the products, including instrumental errors, environmental conditions, and site-specific features [51].

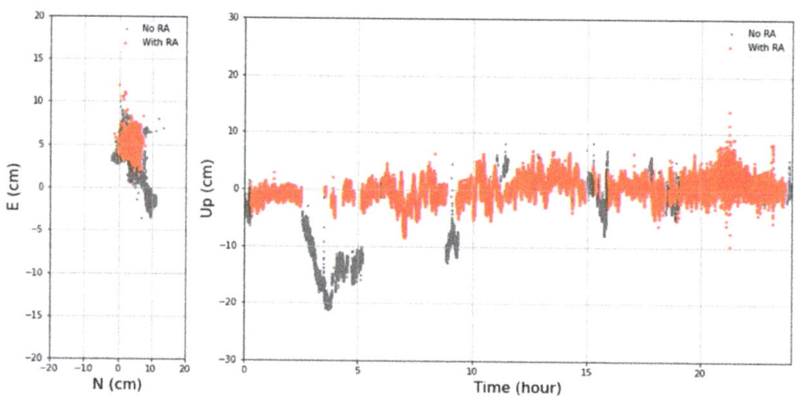

Figure 12. Real-time horizontal (**left**) and vertical (**right**) positioning errors at the MTMR station on 24 January 2024. The GNSS PPP-RA solution (red dots) shows higher precision compared to GNSS PPP epochs (gray dots).

To evaluate the performance of the RING network, daily average RMS values were considered to identify the best and worst stations, as well as their spatial distribution. The classification provides robust information, representing a starting point for future studies with the aim of identifying factors that affect data acquisition and transmission, ultimately improving the overall network performance. The results confirm the capability of real-time PPP for rapidly delivering a centimeter-level positioning accuracy. Over the 22-day investigation period, RING stations exhibited accuracies of within a few centimeters, with significant improvements observed for the full GNSS configuration. The current approach relies on a Continuously Operating Reference Station (CORS) network to enable high-accuracy, fast-converging, and real-time positioning for both GPS-only and full GNSS configurations. The locations of the stations, strategically placed along the Apennine chain or the coast, make the RING network a valuable resource for geodetic data supporting EEW and TEW systems, similar to those developed by researchers in other countries, such as the USA, Chile, or Japan.

5. Conclusions

The preliminary results presented in this study demonstrate the effectiveness of the RING network as a research infrastructure for real-time PPP data analysis. The application of Regional Augmentation within the RTPPP environment further enhances the accuracy of real-time products, even when full GNSS configurations are unavailable. However, multi-frequency and multi-GNSS observations can further improve PPP-RA performance, as shown in the network-wide maps. The accuracies achieved, particularly through short-term window analyses, are particularly relevant for seismic applications, for example, for detecting co-seismic displacements in low-to-moderate-magnitude events (M6) in the Mediterranean Basin. These results are valuable from an interdisciplinary perspective, as geodetic data from GNSS methods can support practical applications and experts in other research fields.

Supplementary Materials: The following supporting information can be downloaded at https://www.mdpi.com/article/10.3390/rs17223661/s1, Figure S1: Histograms showing the distribution of convergence times for the different positioning modes considered in this study: GPS PPP, GNSS PPP, GPS PPP-RA, and GNSS PPP-RA. These histograms represent the time required to achieve positioning errors below 5 and 10 cm for each acquisition gap that occurred during the analyzed time span, taking into account all the RING stations examined in this paper. Most of the RING stations exhibit very short convergence times (a few minutes), particularly when Regional Augmentation is applied, for both GPS-only and full GNSS datasets. However, for some samples (i.e., gaps affecting specific RING sites), convergence times can reach several tens of minutes.

Author Contributions: Conceptualization, P.M. and A.A.; methodology, P.M. and A.A.; software, P.M., M.G. and S.D.; validation, P.M. and A.A.; formal analysis, P.M., A.A. and S.D.; resources, A.A. and M.G.; data curation, P.M. and A.A.; writing—original draft preparation, P.M. and A.A.; writing—review and editing, P.M., A.A., L.F., C.D., S.D., M.G., R.D., N.A.F., C.G., G.P., R.M. and A.V.; visualization, P.M., A.A. and R.M.; supervision, A.A.; project administration, A.A.; funding acquisition, A.A. and M.G. All authors have read and agreed to the published version of the manuscript.

Funding: This research is part of the EWRICA and CIR01-00013 GRINT project, funded by the Italian Ministry of Research.

Data Availability Statement: Data acquired by the RING infrastructure network are available at https://webring.gm.ingv.it/.

Acknowledgments: The authors would like to thank the RING INGV working group for their ongoing efforts and technical support. We also thank GFZ researchers for providing the software used to perform the real-time analysis.

Conflicts of Interest: The authors declare no conflicts of interest.

References

1. Torge, W.; Müller, J.; Pail, R. *Geodesy*; De Gruyter: Berlin, Germany, 2023; ISBN 978-3-11-072329-8.
2. Nikolaidis, R.M.; Bock, Y.; de Jonge, P.J.; Shearer, P.; Agnew, D.C.; Van Domselaar, M. Seismic Wave Observations with the Global Positioning System. *J. Geophys. Res. Solid Earth* **2001**, *106*, 21897–21916. [CrossRef]
3. Larson, K.M.; Bodin, P.; Gomberg, J. Using 1-Hz GPS Data to Measure Deformations Caused by the Denali Fault Earthquake. *Science* **2003**, *300*, 1421–1424. [CrossRef]
4. Bock, Y.; Melgar, D.; Crowell, B.W. Real-Time Strong-Motion Broadband Displacements from Collocated GPS and Accelerometers. *Bull. Seismol. Soc. Am.* **2011**, *101*, 2904–2925. [CrossRef]
5. Melgar, D.; Melbourne, T.I.; Crowell, B.W.; Geng, J.; Szeliga, W.; Scrivner, C.; Santillan, M.; Goldberg, D.E. Real-Time High-Rate GNSS Displacements: Performance Demonstration during the 2019 Ridgecrest, California, Earthquakes. *Seismol. Res. Lett.* **2019**, *91*, 1943–1951. [CrossRef]
6. Boore, D.M.; Bommer, J.J. Processing of Strong-Motion Accelerograms: Needs, Options and Consequences. *Soil Dyn. Earthq. Eng.* **2005**, *25*, 93–115. [CrossRef]

7. Zumberge, J.F.; Heflin, M.B.; Jefferson, D.C.; Watkins, M.M.; Webb, F.H. Precise Point Positioning for the Efficient and Robust Analysis of GPS Data from Large Networks. *J. Geophys. Res. Solid Earth* **1997**, *102*, 5005–5017. [CrossRef]
8. Crowell, B.W.; Melgar, D.; Bock, Y.; Haase, J.S.; Geng, J. Earthquake Magnitude Scaling Using Seismogeodetic Data. *Geophys. Res. Lett.* **2013**, *40*, 6089–6094. [CrossRef]
9. Melgar, D.; Crowell, B.W.; Geng, J.; Allen, R.M.; Bock, Y.; Riquelme, S.; Hill, E.M.; Protti, M.; Ganas, A. Earthquake Magnitude Calculation without Saturation from the Scaling of Peak Ground Displacement. *Geophys. Res. Lett.* **2015**, *42*, 5197–5205. [CrossRef]
10. O'Toole, T.B.; Valentine, A.P.; Woodhouse, J.H. Centroid–Moment Tensor Inversions Using High-Rate GPS Waveforms. *Geophys. J. Int.* **2012**, *191*, 257–270. [CrossRef]
11. Riquelme, S.; Bravo, F.; Melgar, D.; Benavente, R.; Geng, J.; Barrientos, S.; Campos, J. W Phase Source Inversion Using High-Rate Regional GPS Data for Large Earthquakes. *Geophys. Res. Lett.* **2016**, *43*, 3178–3185. [CrossRef]
12. Chen, K.; Liu, Z.; Song, Y.T. Automated GNSS and Teleseismic Earthquake Inversion (AutoQuake Inversion) for Tsunami Early Warning: Retrospective and Real-Time Results. *Pure Appl. Geophys.* **2020**, *177*, 1403–1423. [CrossRef]
13. Kawamoto, S.; Hiyama, Y.; Ohta, Y.; Nishimura, T. First Result from the GEONET Real-Time Analysis System (REGARD): The Case of the 2016 Kumamoto Earthquakes. *Earth Planets Space* **2016**, *68*, 190. [CrossRef]
14. Crowell, B.W.; Melgar, D.; Geng, J. Hypothetical Real-Time GNSS Modeling of the 2016 Mw 7.8 Kaikōura Earthquake: Perspectives from Ground Motion and Tsunami Inundation Prediction. *Bull. Seismol. Soc. Am.* **2018**, *108*, 1736–1745. [CrossRef]
15. Murray, J.R.; Bartlow, N.; Bock, Y.; Brooks, B.A.; Foster, J.; Freymueller, J.; Hammond, W.C.; Hodgkinson, K.; Johanson, I.; López--Venegas, A.; et al. Regional Global Navigation Satellite System Networks for Crustal Deformation Monitoring. *Seismol. Res. Lett.* **2019**, *91*, 552–572. [CrossRef]
16. Crowell, B.W. Chapter 6—Earthquake and Tsunami Early Warning with GNSS Data. In *GNSS Monitoring of the Terrestrial Environment*; Aoki, Y., Kreemer, C., Eds.; Elsevier: Amsterdam, The Netherlands, 2024; pp. 111–127. ISBN 978-0-323-95507-2.
17. Collins, P.; Lahaye, F.; Heroux, P.; Bisnath, S. Precise Point Positioning with Ambiguity Resolution Using the Decoupled Clock Model. In Proceedings of the 21st international technical meeting of the satellite division of the Institute of Navigation, Savannah, Georgia, 16–19 September 2008; Volume 263, pp. 1315–1322.
18. Laurichesse, D.; Mercier, F.; Berthias, J.-P.; Broca, P.; Cerri, L. Integer Ambiguity Resolution on Undifferenced GPS Phase Measurements and Its Application to PPP and Satellite Precise Orbit Determination. *Navigation* **2009**, *56*, 135–149. [CrossRef]
19. Cai, C.; Gao, Y. Modeling and Assessment of Combined GPS/GLONASS Precise Point Positioning. *GPS Solut.* **2013**, *17*, 223–236. [CrossRef]
20. Yigit, C.O.; Gikas, V.; Alcay, S.; Ceylan, A. Performance Evaluation of Short to Long Term GPS, GLONASS and GPS/GLONASS Post-Processed PPP. *Surv. Rev.* **2014**, *46*, 155–166. [CrossRef]
21. Cai, C.; Gao, Y.; Pan, L.; Zhu, J. Precise Point Positioning with Quad-Constellations: GPS, BeiDou, GLONASS and Galileo. *Adv. Space Res.* **2015**, *56*, 133–143. [CrossRef]
22. Tsushima, H.; Ohta, Y. Review on Near-Field Tsunami Forecasting from Offshore Tsunami Data and Onshore GNSS Data for Tsunami Early Warning. *J. Disaster Res.* **2014**, *9*, 339–357. [CrossRef]
23. Melgar, D.; Allen, R.M.; Riquelme, S.; Geng, J.; Bravo, F.; Baez, J.C.; Parra, H.; Barrientos, S.; Fang, P.; Bock, Y.; et al. Local Tsunami Warnings: Perspectives from Recent Large Events. *Geophys. Res. Lett.* **2016**, *43*, 1109–1117. [CrossRef]
24. Báez, J.C.; Leyton, F.; Troncoso, C.; del Campo, F.; Bevis, M.; Vigny, C.; Moreno, M.; Simons, M.; Kendrick, E.; Parra, H.; et al. The Chilean GNSS Network: Current Status and Progress toward Early Warning Applications. *Seismol. Res. Lett.* **2018**, *89*, 1546–1554. [CrossRef]
25. Jamil, H.; Mohamed, D.A.; Chang, D. The Malaysia Real-Time Kinematic GNSS Network (MyRTKnet) in 2010 and Beyond. In Proceedings of the FIG Congress 2010, Sydney, Australia, 11–16 April 2010.
26. Hodgkinson, K.M.; Mencin, D.J.; Feaux, K.; Sievers, C.; Mattioli, G.S. Evaluation of Earthquake Magnitude Estimation and Event Detection Thresholds for Real-Time GNSS Networks: Examples from Recent Events Captured by the Network of the Americas. *Seismol. Res. Lett.* **2020**, *91*, 1628–1645. [CrossRef]
27. Chousianitis, K.; Papanikolaou, X.; Drakatos, G.; Tselentis, G. Akis NOANET: A Continuously Operating GNSS Network for Solid--Earth Sciences in Greece. *Seismol. Res. Lett.* **2021**, *92*, 2050–2064. [CrossRef]
28. Avallone, A.; Selvaggi, G.; D'Anastasio, E.; D'Agostino, N.; Pietrantonio, G.; Riguzzi, F.; Serpelloni, E.; Anzidei, M.; Casula, G.; Cecere, G.; et al. The RING Network: Improvement of a GPS Velocity Field in the Central Mediterranean. *Ann. Geophys.* **2010**, *53*, 39–54. [CrossRef]
29. Devoti, R.; Pietrantonio, G.; Riguzzi, F. GNSS Networks for Geodynamics in Italy. *Fisica Tierra* **2014**, *26*, 11–24. [CrossRef]
30. Chiarabba, C.; De Gori, P.; Chiaraluce, L.; Bordoni, P.; Cattaneo, M.; De Martin, M.; Frepoli, A.; Michelini, A.; Monachesi, A.; Moretti, M.; et al. Mainshocks and Aftershocks of the 2002 Molise Seismic Sequence, Southern Italy. *J. Seismol.* **2005**, *9*, 487–494. [CrossRef]

31. Serpelloni, E.; Anderlini, L.; Belardinelli, M.E. Fault Geometry, Coseismic-Slip Distribution and Coulomb Stress Change Associated with the 2009 April 6, Mw 6.3, L'Aquila Earthquake from Inversion of GPS Displacements. *Geophys. J. Int.* **2012**, *188*, 473–489. [CrossRef]
32. Cheloni, D.; De Novellis, V.; Albano, M.; Antonioli, A.; Anzidei, M.; Atzori, S.; Avallone, A.; Bignami, C.; Bonano, M.; Calcaterra, S.; et al. Geodetic Model of the 2016 Central Italy Earthquake Sequence Inferred from InSAR and GPS Data. *Geophys. Res. Lett.* **2017**, *44*, 6778–6787. [CrossRef]
33. Melgar, D.; Crowell, B.W.; Bock, Y.; Haase, J.S. Rapid Modeling of the 2011 Mw 9.0 Tohoku-Oki Earthquake with Seismogeodesy. *Geophys. Res. Lett.* **2013**, *40*, 2963–2968. [CrossRef]
34. Alkan, R.M.; Erol, S.; Ozulu, I.M.; Ilci, V. Accuracy Comparison of Post-Processed PPP and Real-Time Absolute Positioning Techniques. *Geomat. Nat. Hazards Risk* **2020**, *11*, 178–190. [CrossRef]
35. Wabbena, G.; Schmitz, M.; Bagge, A. PPP-RTK: Precise Point Positioning Using State-Space Representation in RTK Networks. In Proceedings of the 18th international technical meeting of the satellite division of the Institute of navigation, Long Beach, CA, USA, 13–16 September 2005; Volume 259, pp. 2584–2594.
36. Geng, J.; Teferle, F.N.; Meng, X.; Dodson, A.H. Towards PPP-RTK: Ambiguity Resolution in Real-Time Precise Point Positioning. *Adv. Space Res.* **2011**, *47*, 1664–1673. [CrossRef]
37. Teunissen, P.J.G.; Odijk, D.; Zhang, B. PPP-RTK: Results of CORS Network-Based PPP with Integer Ambiguity Resolution. *J. Aeronaut. Astronaut. Aviat. Ser. A* **2010**, *42*, 223–230.
38. Zhang, X.; Li, X.; Guo, F. Satellite Clock Estimation at 1 Hz for Realtime Kinematic PPP Applications. *GPS Solut.* **2011**, *15*, 315–324. [CrossRef]
39. Li, X.; Ge, M.; Dai, X.; Ren, X.; Fritsche, M.; Wickert, J.; Schuh, H. Accuracy and Reliability of Multi-GNSS Real-Time Precise Positioning: GPS, GLONASS, BeiDou, and Galileo. *J. Geod.* **2015**, *89*, 607–635. [CrossRef]
40. Amato, A.; Mele, F. Performance of the INGV National Seismic Network from 1997 to 2007. *Ann. Geophys.* **2008**, *51*, 417–431. [CrossRef]
41. D'Agostino, N.; Avallone, A.; Cheloni, D.; D'Anastasio, E.; Mantenuto, S.; Selvaggi, G. Active Tectonics of the Adriatic Region from GPS and Earthquake Slip Vectors. *J. Geophys. Res. Solid Earth* **2008**, *113*, B12413. [CrossRef]
42. Chiarabba, C.; De Gori, P.; Mele, F.M. Recent Seismicity of Italy: Active Tectonics of the Central Mediterranean Region and Seismicity Rate Changes after the Mw 6.3 L'Aquila Earthquake. *Tectonophysics* **2015**, *638*, 82–93. [CrossRef]
43. Serpelloni, E.; Cavaliere, A.; Martelli, L.; Pintori, F.; Anderlini, L.; Borghi, A.; Randazzo, D.; Bruni, S.; Devoti, R.; Perfetti, P.; et al. Surface Velocities and Strain-Rates in the Euro-Mediterranean Region From Massive GPS Data Processing. *Front. Earth Sci.* **2022**, *10*, 907897. [CrossRef]
44. Falco, L. Implementazione e Gestione di una Rete di Monitoraggio GPS e Sismica Mediante Tecnologie GPRS/EDGE/UMTS/HSDPA. 2008. Available online: https://editoria.ingv.it/archivio_pdf/rapporti/68/pdf/rapporti_69.pdf (accessed on 10 June 2025).
45. Margheriti, L.; Nostro, C.; Cocina, O.; Castellano, M.; Moretti, M.; Lauciani, V.; Quintiliani, M.; Bono, A.; Mele, F.M.; Pintore, S.; et al. Seismic Surveillance and Earthquake Monitoring in Italy. *Seismol. Res. Lett.* **2021**, *92*, 1659–1671. [CrossRef]
46. Devoti, R.; D'Agostino, N.; Serpelloni, E.; Pietrantonio, G.; Riguzzi, F.; Avallone, A.; Cavaliere, A.; Cheloni, D.; Cecere, G.; D'Ambrosio, C.; et al. A Combined Velocity Field of the Mediterranean Region. *Ann. Geophys.* **2017**, *60*, S0215. [CrossRef]
47. Malys, S.; Jensen, P.A. Geodetic Point Positioning with GPS (Global Positioning System) Carrier Beat Phase Data from the CASA (Central and South America) Uno Experiment. *Geophys. Res. Lett.* **1990**, *17*, 5. [CrossRef]
48. Li, X.; Wang, B.; Li, X.; Huang, J.; Lyu, H.; Han, X. Principle and Performance of Multi-Frequency and Multi-GNSS PPP-RTK. *Satell. Navig.* **2022**, *3*, 7. [CrossRef]
49. Ge, M.; Douša, J.; Li, X.; Ramatschi, M.; Nischan, T.; Wickert, J. A Novel Real-Time Precise Positioning Service System: Global Precise Point Positioning With Regional Augmentation. *J. Glob. Position. Syst.* **2012**, *11*, 2–10. [CrossRef]
50. Avallone, A.; Marzario, M.; Cirella, A.; Piatanesi, A.; Rovelli, A.; Di Alessandro, C.; D'Anastasio, E.; D'Agostino, N.; Giuliani, R.; Mattone, M. Very High Rate (10 Hz) GPS Seismology for Moderate-Magnitude Earthquakes: The Case of the Mw 6.3 L'Aquila (Central Italy) Event. *J. Geophys. Res. Solid Earth* **2011**, *116*, B02305. [CrossRef]
51. Karaim, M.; Elsheikh, M.; Noureldin, A.; Karaim, M.; Elsheikh, M.; Noureldin, A. GNSS Error Sources. In *Multifunctional Operation and Application of GPS*; IntechOpen: London, UK, 2018; ISBN 978-1-78923-215-8.

remote sensing

MDPI

Article

Assessing the Performance of GNSS-IR for Sea Level Monitoring During Hurricane-Induced Storm Surges

Runtao Zhang [1], Kai Liu [2,3,*], Xue Wang [4], Zhao Li [2], Tao Xie [5], Qusen Chen [2] and Xin Chang [1]

1 School of Geodesy and Geomatics, Wuhan University, Wuhan 430079, China; 2022302141004@whu.edu.cn (R.Z.); xchang@sgg.whu.edu.cn (X.C.)
2 GNSS Research Center, Wuhan University, Wuhan 430079, China; zhao.li@whu.edu.cn (Z.L.); chenqs@whu.edu.cn (Q.C.)
3 State Key Laboratory of Precision Geodesy, Innovation Academy for Precision Measurement Science and Technology, CAS, Wuhan 430079, China
4 Beijing FreeDo Technology Co., Ltd., Beijing 100176, China; wangxue@freedotech.com
5 The Third Surveying and Mapping Institute of Guizhou Province, Guiyang 550004, China; zbss_xt@163.com
* Correspondence: whu_lk@whu.edu.cn

Abstract: With the intensification of extreme climate change, hurricanes are becoming increasingly frequent, and coastal regions are often impacted by hurricane-induced storm surges. While GNSS-IR (Global Navigation Satellite System–Interferometric Reflectometry) has been widely used for sea level monitoring, its application in extreme weather events such as storm surges remains limited. This study focuses on GNSS-IR-based storm surge monitoring and investigates six hurricane events using data from two GNSS stations (CALC and FLCK) located in the Gulf of Mexico. The monitoring accuracy and effectiveness are systematically evaluated. Results indicate that GNSS-IR achieves a sea level accuracy of approximately 7 cm under non-storm surge conditions. Compared with the FLCK station, the CALC station has a wider field of water reflection and higher precision observation results. This further confirms that an open environment is a prerequisite for ensuring the accuracy of GNSS-IR measurements. However, accuracy degrades significantly during storm surges, reaching only a decimeter-level precision. Multi-GNSS observations notably improve temporal resolution, with valid observation periods covering 83% to 97% of the total time, compared with only 40% to 60% for single-system observations. Moreover, dynamic sea level variations are closely correlated with hurricane trajectories, which affects GNSS-IR measurement accuracy to some extent. The GPS L2 band is particularly sensitive, likely due to the complex surface-reflected condition caused by hurricanes. Despite reduced accuracy during storm surges, GNSS-IR remains capable of capturing dynamic sea level changes effectively, demonstrating its potential as a valuable supplement to the existing observation networks for extreme weather monitoring.

Keywords: GNSS-IR; hurricane; sea level; storm surge; wind speed

1. Introduction

Storm surges are characterized by significant abnormal rises in sea level, primarily driven by strong cyclonic systems such as hurricanes, under the combined influence of low atmospheric pressure and intense wind fields [1]. These extreme events are often accompanied by heavy rainfall, high waves, and coastal flooding, posing severe threats to coastal infrastructure, ecosystems, and human life and property. In 2005, Hurricane Katrina made landfall in Louisiana, U.S., causing extensive damage and displacing approximately 657,000 people due to flooding [2]. In August 2020, Hurricane Laura, a Category 4 storm, struck southwestern Louisiana, resulting in at least 77 fatalities and over USD 10 billion in economic losses [3]. Against the backdrop of global climate change and ongoing relative sea level rise, low-lying coastal regions around the world are expected to face more frequent and destructive storm surges in the coming decades, potentially placing hundreds of millions of people at risk [4,5]. The continuous, accurate, and reliable monitoring of anomalous sea level variations during storm surges is not only a fundamental observational basis for current storm surge research, but also a critical prerequisite for analyzing and assessing their hazard mechanisms and impacts.

Conventional sea level monitoring techniques primarily include tide gauges and satellite radar altimetry. Tide gauges can provide high-accuracy measurements of sea surface height, but their construction and maintenance are expensive, hindering large-scale deployment. Moreover, traditional tide gauges are typically installed near the waterline and are susceptible to structural damage or operational failures during extreme weather events such as storm surges, often resulting in data interruptions [6–8]. While satellite radar altimetry offers high measurement accuracy and broad spatial coverage over open oceans, its performance near coastlines is substantially degraded, and it struggles to capture the rapid changes in sea surface height that occur during storm surge events [9,10]. A single measurement technique inevitably has inherent limitations and cannot fully reveal the characteristics of sea level changes at different spatial and temporal scales.

GNSS is currently widely used in earth science research, including crustal deformation, disaster monitoring, hydrological assessment, and many other fields [11]. The GNSS Interferometric Reflectometry (GNSS-IR) technique was first introduced by Larson in 2008, offering a novel approach for sea level monitoring by utilizing GNSS signals reflected from the water surface [12]. The GNSS-IR technique determines water level variations by analyzing the Signal-to-Noise Ratio (SNR) signals to calculate the vertical distance from the reflecting surface to the antenna phase center [13]. Löfgren et al. (2014) improved the inversion method by incorporating tidal models based on data from five global sites and investigated the characteristics and patterns of long-term sea level change [14]. Since then, researchers have expanded the application of GNSS-IR to various water bodies, including seas [15], rivers [16], lakes [17], and reservoirs [18]. These results demonstrate that GNSS-IR can achieve a high retrieval accuracy under stable water level conditions. Under different water conditions, there are significant differences in measurement accuracy. For example, in calm water bodies such as reservoirs, the GNSS can achieve millimeter-level accuracy in long-term, water level observation sequences [19]. In pumped-storage power stations, when water levels change very rapidly, accuracy can only be maintained at around 30 cm [20]. In sea level monitoring, due to varying tidal movement amplitudes among different stations, measurement accuracy ranges from the centimeter level to decimeter level [21,22]. Collectively, these studies indicate that factors such as sea surface fluctuations, their amplitude, and rate of change can all influence the GNSS-IR measurement accuracy. However, most existing studies have focused on calm hydrological conditions, while under extreme events such as typhoons, tsunamis, and storm surges, the accuracy of the GNSS-IR inevitably faces substantially different challenges.

In recent years, the application of GNSS-IR in hydrological monitoring has continued to expand, and researchers have progressively carried out experiments and evaluations on water level monitoring under extreme conditions. Peng et al. (2019) used GPS data at the HKQT and HKPC stations in Hong Kong to retrieve the sea surface height during a storm surge event [23]. Since then, numerous studies have further investigated the applicability of GNSS-IR under extreme sea level conditions across various regions [24–26], progressively expanding its scope of application. Multi-GNSS GNSS-IR studies have demonstrated that the technique can effectively capture abnormal sea level variations caused by events such as storm surges and tsunamis, serving as a valuable supplement to traditional coastal hazard early warning systems [27]. Owing to the widespread global distribution of GNSS stations, integrating observations from multiple stations enables the GNSS-IR to detect spatial variations in sea level within the same region during storm surges, which overcomes the spatial limitations of conventional tide gauges [28]. Moreover, GNSS-IR can be utilized for tsunami inference, offering a potential alternative to traditional seafloor monitoring equipment [29]. These findings indicate the high adaptability of GNSS-IR technology during extreme sea level events. However, these studies were conducted at different sites with varying measurement accuracies and storm surge intensities, making it difficult to comprehensively assess how GNSS-IR accuracy responds to extreme water level changes. Results based on a single site or event may be influenced by case-specific variability and thus lack broader representativeness. Although previous research has preliminarily confirmed the feasibility of GNSS-IR during storm surge events, a systematic evaluation of its stability remains lacking—particularly with respect to its applicability and robustness under varying storm surge intensities, tracks, and site-specific environmental conditions.

To analyze the accuracy and reliability of GNSS-IR in monitoring abnormal water levels, this study selected two continuously operating GNSS stations (CORS) located along the Gulf of Mexico coast in the United States. By analyzing six typical storm surge events between 2017 and 2024, we assessed the monitoring capabilities and robustness of GNSS-IR under extreme conditions in the Gulf region and comprehensively analyzed the spatiotemporal evolution of water level anomalies caused by hurricane movements. This study systematically evaluates the sea level monitoring performance and accuracy of GNSS-IR under varying oceanic conditions and further explores its potential applications in monitoring and assessing extreme climate events.

2. Data and Methods

2.1. Study Area and GNSS Datasets

This study selects two nearshore GNSS stations with favorable observation environments and GNSS-IR reflection conditions: the CALC (ID: 10121) station located on the southern coast of Louisiana, U.S. and the FLCK (ID: 10411) station along the central coast of Florida, U.S. (Figure 1). For the CALC station, it equipped with a Trimble Alloy GNSS receiver (Trimble Inc., Sunnyvale, CA, USA), and a Trimble TRM115000.00 antenna (Trimble Inc., Sunnyvale, CA, USA). For the FLCK station, it equipped with a Leica GR50 GNSS receiver (Leica Geosystems AG, Heerbrugg, Switzerland), and a LEIAR20 LEIM antenna (Leica Geosystems AG, Heerbrugg, Switzerland). The CALC station is surrounded by open terrain with no obstructions, and its proximity to the sea provides ideal GNSS reflection surfaces. Therefore, GNSS data selection for this site is limited to an azimuth range of 0–360° and an elevation angle range of 10–30°. CALC is a continuously operating GNSS reference station jointly established by the U.S. National Oceanic and Atmospheric Administration (NOAA) and UNAVCO. The FLCK station is located near some non-water reflective surfaces such as docks. To reduce interference from these structures, the GNSS data selection at FLCK is restricted to an azimuth range of 60–330° and an elevation angle range of 5–30°. These two stations can receive signals from GPS, GLONASS, and Galileo systems. Both stations are co-located with NOAA tide gauges (CALC: Station ID 8768094, FLCK: Station ID 8727520), which provide measured tide level data and wind speed observations at six-minute intervals. The tide gauge data were obtained from the official NOAA Tides & Currents website (https://tidesandcurrents.noaa.gov/stationhome.html?id=8768094 for the CALC station and https://tidesandcurrents.noaa.gov/stationhome.html?id=8727520 for the FLCK station). The vertical antenna heights relative to the long-term mean sea level are 12.6 m for CALC and 10.2 m for FLCK, and the GNSS observation data have a sampling interval of 30 s.

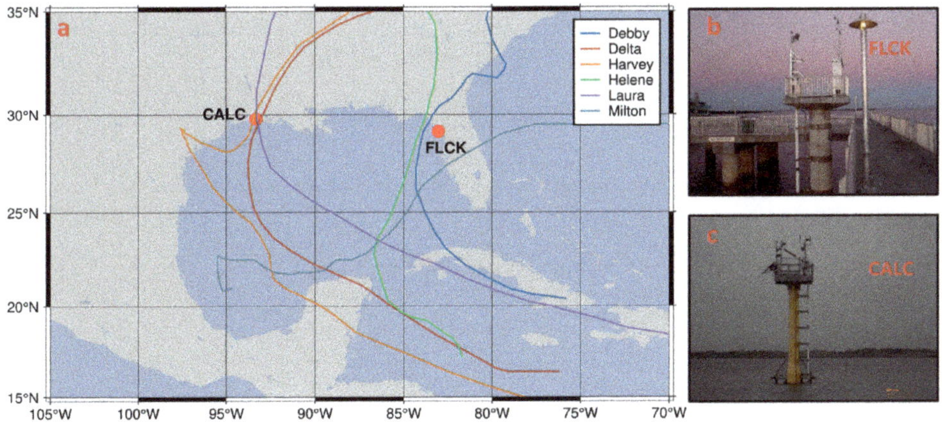

Figure 1. Station distributions and hurricane tracks of the experiment in Figure (**a**) and photographs of CALC and FLCK in Figure (**b**,**c**).

2.2. Hurricane Datasets

To systematically evaluate the adaptability and robustness of GNSS-IR technology under storm surge conditions, this study selected six representative hurricane events occurring between 2017 and 2024, with each station experiencing three storm surge events. Hurricane information was obtained from the NOAA National Hurricane Center (https://www.nhc.noaa.gov (accessed on 26 July 2025)), and the detailed overview is presented in Table 1. Figure 1 summarizes the trajectories of these six hurricanes to illustrate their spatial relationship with the GNSS stations, along with onsite photographs of the two stations (image source: https://geodesy.noaa.gov (accessed on 26 July 2025)). These six hurricane events vary significantly in terms of spatial distribution (Figure 1), with maximum wind speeds ranging from 130 km/h to 286 km/h. This significant variability ensures a more comprehensive and objective testing and evaluation study.

Table 1. General information on the six hurricanes provided by NOAA.

Hurricane	Time	Maximum Wind Speed	The Closest Distance to the Station
Laura	August 2020	240 km/h	<10 km (CALC)
Delta	October 2020	222 km/h	≈30 km (CALC)
Harvey	August 2017	213 km/h	<20 km (CALC)
Debby	August 2024	130 km/h	≈80 km (FLCK)
Helene	September 2024	222 km/h	≈110 km (FLCK)
Milton	October 2024	286 km/h	≈200 km (FLCK)

2.3. Methods

The GNSS receiver simultaneously receive direct signals from satellites as well as indirect signals reflected off surfaces such as the ground and water. These two types of signals interfere and superimpose at the receiver, resulting in multipath effects [12]. This phenomenon is illustrated in Figure 2. The multipath interference effect manifests in GNSS observation data as periodic oscillations in the Signal-to-Noise Ratio (SNR). Since the direct signal strength is typically much stronger than that of the reflected signal, a low-order polynomial is commonly fitted to the original SNR data to model and remove the trend component. In this study, a uniform second-order polynomial was applied to obtain the detrended SNR (DSNR). This process yields the DSNR, which facilitates the effective extraction and analysis of the periodic features associated with the reflected signals. After detrending, the DSNR exhibits clear periodic oscillations [30]:

$$\mathrm{DSNR} = 2A_d A_m \cos\psi = A\cos\left(\frac{2\pi}{\lambda}\cdot 2h\sin\theta' + \psi_0\right) \tag{1}$$

where θ' is the satellite elevation angle corrected for tropospheric delay using the Ulich model [31]:

$$\Delta\theta = 10^{-6} N \frac{\cos\theta}{\sin\theta + 0.00175\tan(87.5-\theta)},\ \theta' = \theta + \Delta\theta \tag{2}$$

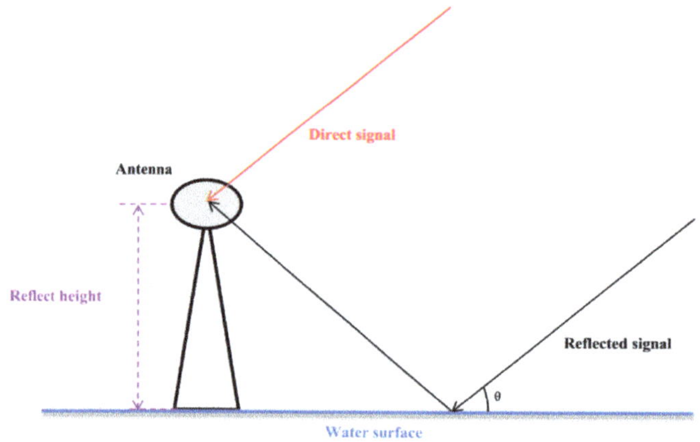

Figure 2. The GNSS direct and reflected signals are received by the antenna and undergo interference.

By defining $\mathrm{t} = 2\sin\theta'$, $f = 2h/\lambda$, the equation can be rewritten as

$$\mathrm{DSNR} = A\cos(\pi\cdot ft + \psi_0) \tag{3}$$

Equations (1) and (3) indicate that the oscillation frequency of the reflected signal is directly related to the effective reflection height h, defined as the vertical distance from the reflecting surface to the phase center of the GNSS antenna. This relationship is expressed as $f = 2h/\lambda$, where λ is the signal wavelength.

To extract the dominant frequency f, spectral analysis of the DSNR was performed using the Lomb–Scargle Periodogram. The high-frequency cutoff was dynamically estimated based on the Nyquist frequency, determined by the satellite arc length and the number of samples within the arc, and further scaled by a prescribed maximum reflector height to avoid an excessively wide search range. To exclude spurious peaks while retaining as much data as possible during storm surge conditions, a minimum peak-to-noise ratio threshold of 2 was imposed. Additionally, candidate peaks were restricted to the range of the long-term relative mean sea level at each station (CALC: 12.6 m, FLCK: 10.2 m). During calm periods, this range was set to ±2.5 m, while during storm surge conditions, it was expanded to ±5 m.

We evaluate the feasibility of GNSS-IR for storm surge monitoring and examine its retrieval accuracy during such events. Two coastal GNSS stations on the Gulf of Mexico—CALC and FLCK—were selected for a series of retrieval experiments covering multiple representative storm surges from 2017 to 2024. For the CALC station, signals within an elevation angle range of 10–30° and an azimuth range of 0–360° were utilized, whereas for FLCK the elevation angle range was set to 5–30° and the azimuth range to 60–330°. To ensure the robustness of these estimates, we applied a three times median absolute deviation (3 MAD) filter and performed outlier rejection using a 4 h moving window. Multi-GNSS integration was performed by simply concatenating results from different constellations without weighting to maximize temporal coverage. This quality control procedure yielded a continuous GNSS-IR sea surface height time series. The retrieved heights were then compared against contemporaneous tide gauge measurements to assess both accuracy and trend consistency.

3. Results of GNSS-IR-Based Storm Surge Monitoring

3.1. Identification and Monitoring of Storm Surge Events

Figures 3 and 4 present a comparison between GNSS-IR-retrieved sea surface height (SSH) retrievals and tide gauge measurements across multiple storm surge events. Except for a data gap at FLCK on DOY 273–274 (2024), the dataset is complete. Overall, the results demonstrate that GNSS-IR effectively captures rapid water level fluctuations associated with storm surge episodes. During relatively calm periods, water level variations are slow and consistent. Under such conditions, all three satellite constellations—GPS, GLONASS, and Galileo—can effectively capture key oscillation features. The retrieved sea surface heights show a strong agreement with tide gauge data, indicating that GNSS-IR enables high-accuracy, continuous water level monitoring under stable meteorological conditions.

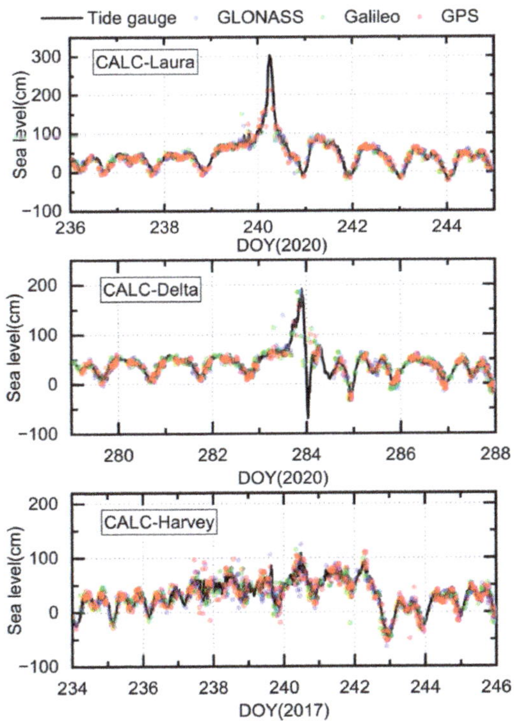

Figure 3. Comparison between GNSS-IR-retrieved sea surface heights and tide gauge observations at the CALC station during hurricanes Laura, Delta, and Harvey. The black lines represent tide gauge measurements, while the colored dots denote GNSS-IR retrievals.

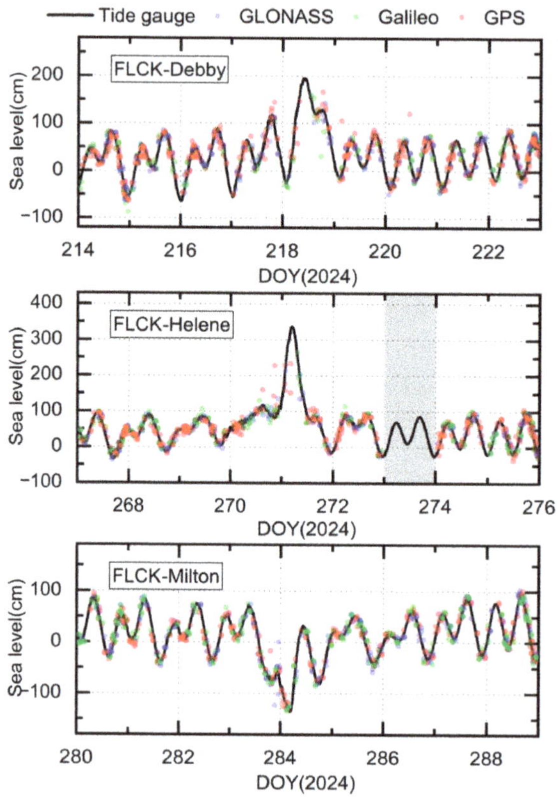

Figure 4. Comparison between GNSS-IR-retrieved sea surface heights and tide gauge observations at the FLCK station during hurricanes Debby, Helene, and Milton. The black lines represent tide gauge measurements, while the colored dots denote GNSS-IR retrievals.

In contrast, during peak storm surge conditions, water levels become highly irregular, rising or falling sharply over short time spans. In these cases, more outliers appear compared to calm periods. Despite the increased overall retrieval error under extreme conditions, most GNSS-IR estimates still remain close to the actual tide levels, confirming the method's ability to track major water level fluctuations even during extreme events. To further evaluate the retrieval performance of different GNSS constellations under storm surge conditions, two representative events—Hurricane Laura and Hurricane Helene—were selected. For each event, a two-day observation window covering the peak storm surge impact was analyzed. Sea surface height retrievals from GPS, GLONASS, and Galileo and the integration of multiple GNSS systems were compared with collocated tide gauge records. The results are shown in Figures 5 and 6.

In Figures 5 and 6, the black solid line represents the observed sea surface heights recorded by NOAA tide gauges, while colored scatter points correspond to GNSS-IR retrievals from different frequency bands across the three satellite constellations. Among the single GNSS systems, GPS contributes the highest number of observations, whereas other systems show relatively limited coverage. This difference is less noticeable under calm conditions but becomes more pronounced during the peak phases of storm surges when significant fluctuations occur and valid observations become sparse. The integration of multiple GNSS systems can partially mitigate this issue by increasing observational density, thereby enhancing the monitoring of abnormal sea level changes. At the CALC station, GPS-based retrievals closely track the entire storm surge process, with only occasional larger deviations. Meanwhile, at the FLCK station, the Galileo system successfully identified an extreme water level peak of approximately 250 cm. All three GNSS systems demonstrate high retrieval accuracy when water levels remain below 150 cm, further confirming the applicability of GNSS-IR under extreme water level conditions. We evaluated the sea level inversion accuracy and numbers of results of individual GNSS systems and single-frequency bands during the main impact periods of two typical storm surge events, Hurricane Laura and Hurricane Helene, and further assessed the overall performance after multi-system integration (Table 2). During the main impact periods of storm surges, the RMSEs of most single-frequency GNSS signals remained below 20 cm, indicating a generally good inversion accuracy. During Hurricane Laura, the three GPS bands at the CALC station demonstrated a stable inversion performance, with RMSEs ranging from 10.39 to 15.18 cm. Especially, both the L1 and L2 bands produced more than 40 valid observations, indicating both high accuracy and strong reliability. The GLONASS system also maintained a high level of accuracy, with RMSEs below 20 cm and slightly lower, but still enough, valid results (over 30 points). Due to the limited number of satellites, the Galileo system yielded fewer inversion results and relatively higher RMSEs. In contrast, during Hurricane Helene, the inversion accuracy at the FLCK station declined, with the GPS L2 band being particularly affected—its RMSE reached 51.59 cm with only 13 valid results, suggesting that the signal quality or reflectivity may have been severely disrupted. Under the same conditions, the GLONASS and Galileo systems still maintained a relatively better inversion performance, with RMSEs of 21.42 cm and 15.53 cm, respectively, and a reasonable number of results. This indicates that different GNSS systems exhibit varying levels of resilience under extreme environmental conditions.

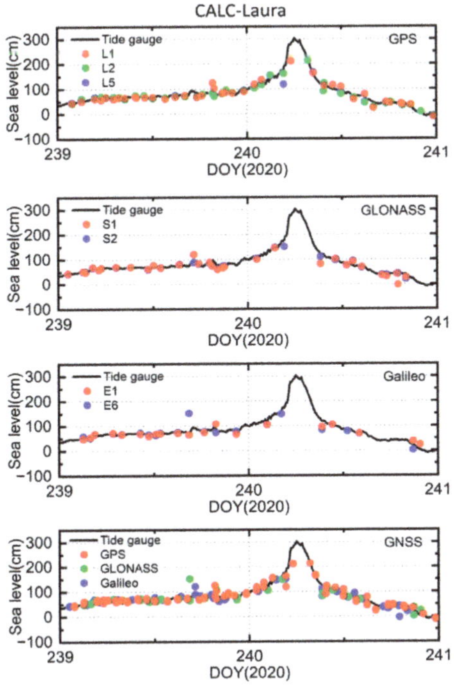

Figure 5. Comparison between GNSS-IR-retrieved sea surface heights from GPS, GLONASS, Galileo, and GNSS and tide gauge measurements at the CALC station during the peak periods (DOY 239–241) of Hurricane Laura.

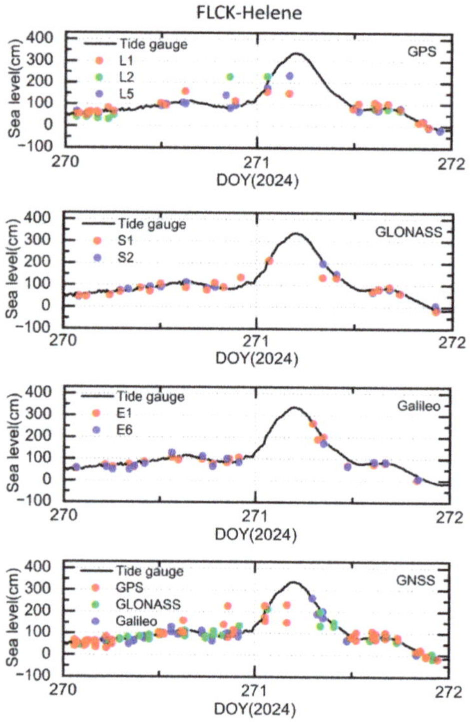

Figure 6. Comparison between GNSS-IR-retrieved sea surface heights from GPS, GLONASS, Galileo, and GNSS and tide gauge measurements at the FLCK station during the peak periods (DOY 270–272) of Hurricane Helene.

Table 2. GNSS-IR sea level inversion accuracy and numbers of results for individual signals and multi-system fusion during the main impact periods of Hurricanes Laura and Helene.

System	Band	Hurricane Laura		Hurricane Helene	
		RMSE (cm)	Num	RMSE (cm)	Num
GPS	L1	14.90	40	39.52	25
	L2	10.39	41	51.59	13
	L5	15.18	22	27.23	20
GLONASS	S1	14.86	31	21.42	20
	S2	8.24	30	7.70	16
Galileo	E1	11.03	20	15.53	16
	E6	19.83	21	13.77	20
GNSS	---	13.61	205	28.59	130

After multi-system fusion, the number of valid GNSS-IR inversion results increased significantly. The fused RMSEs were 13.61 cm (205 points) at CALC and 28.59 cm (130 points) at FLCK. These results indicate that at CALC, fused multi-system GNSS-IR observations can achieve high-precision sea level monitoring. Although accuracy at FLCK was lower, given that water levels during Hurricane Helene exceeded 3 m, an RMSE of 28.59 cm remains within an acceptable range. The results in both accuracy and data volume after fusion demonstrate that multi-system GNSS-IR fusion contributes to more stable and continuous sea level monitoring during storm surge events, showing a strong adaptability and robustness. Overall, GNSS-IR exhibits a robust and transferable performance across different regions and storm surge events. Whether over extended calm periods or short-duration extreme events, it reliably tracks sea level fluctuations, underscoring its practical potential for storm surge monitoring.

3.2. Assessment of GNSS-IR Retrieval Accuracy During Storm Surges

Table 3 presents the root mean square errors (RMSE, cm) of GNSS-IR sea surface height retrievals for GPS, GLONASS, and Galileo across six storm surge events. This comparison highlights notable variations in retrieval accuracy among different GNSS constellations, frequency bands, and monitoring stations during storm surges. Considering differences among systems and frequency bands, the GPS constellation demonstrates a superior overall retrieval performance. Specifically, the L1 frequency band maintains an RMSE of about 10 cm across multiple events, indicating good stability. Although the L2 band achieves high accuracy at the CALC station, significant error increases occur at the FLCK station during the Debby and Milton events. Notably, the RMSE reached 95.26 cm during Hurricane Milton, suggesting that environmental factors may severely degrade L2 band accuracy in certain conditions. The L5 band shows excellent precision and stability in available events, with the lowest RMSE of 5.48 cm observed during the Delta event. Due to Hurricane Harvey occurring in 2017, when the CALC station did not yet support GPS L5 signal reception, L5 data for that period is unavailable. For the GLONASS system, the S1 and S2 bands exhibit a stable performance, with RMSE generally near 10 cm. The S2 band achieved the lowest RMSE in the table—6.06 cm—during the Laura event. In contrast, the Galileo system's E1 and E6 bands have slightly higher errors, with the RMSE mostly between 11 and 14 cm.

Table 3. RMSE (cm) of GNSS-IR sea level retrievals relative to tide gauge observations across different frequency bands and storm surge events.

System	Band	Laura	Delta	Harvey	Debby	Helene	Milton
GPS	L1	7.58	8.35	13.40	14.65	17.94	10.20
	L2	7.20	7.47	11.98	40.60	24.55	95.26
	L5	7.52	5.48	---	11.56	15.70	11.71
GLONASS	S1	7.74	11.16	13.83	12.30	12.02	10.98
	S2	6.06	7.53	9.71	13.59	9.71	13.27
Galileo	E1	9.43	13.07	12.16	13.21	13.00	11.48
	E6	9.60	14.01	11.64	16.43	13.08	12.78
GNSS	---	7.92	10.05	12.25	18.71	15.67	31.61

From a site-specific perspective, the CALC station is located in an open area surrounded by water, providing favorable observation conditions (Figure 1b). As a result, its retrievals consistently show a lower RMSE. In contrast, the FLCK station is located near non-water reflective surfaces such as piers (Figure 1c), which contribute to a higher RMSE during certain storm surge events (e.g., Hurricane Milton). This suggests that the performance of the GNSS-IR at FLCK may be more susceptible to environmental interference. GNSS-IR retrieval accuracy is jointly influenced by the GNSS system, frequency band, station environment, and specific event characteristics. Future studies could explore weighted multi-constellation and multi-frequency fusion strategies that assign appropriate weights to frequency bands based on historical event data at each station, aiming to enhance the stability and reliability of GNSS-IR monitoring under complex storm surge conditions.

4. Discussion

4.1. The Impact of Hurricane Migration

The six storm surge events were categorized into three stages—before, during, and after the storm surge—based on the distance between the hurricane eye and the GNSS station. These stages correspond to the light green, light red, and light blue shaded areas in Figures 7 and 8, respectively. Using this classification, Figures 7 and 8 present a statistical visualization showing how GNSS-IR retrieval deviations vary with sea surface wind speed and the distance from the hurricane eye. In the plots, the black solid line shows the distance between the hurricane eye and the GNSS station, the red curve represents sea surface wind speed, and the blue scatter points indicate the deviation of the GNSS-IR-retrieved sea surface height from tide gauge observations.

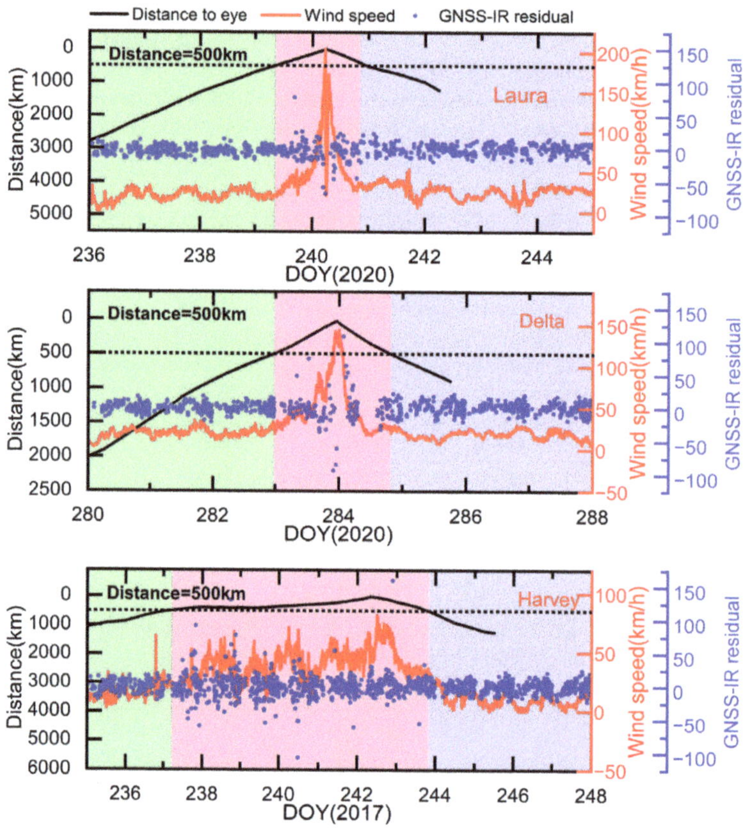

Figure 7. Variations in GNSS-IR retrieval deviations at the CALC station during three storm surge events in relation to wind speed and the distance between the hurricane eye and the station.

During the period before the storm surge, the hurricane remains far from the station, and local wind speeds are relatively low and stable. Under these conditions, GNSS-IR retrievals show limited deviation, with most values confined within ±25 cm, indicating a good overall accuracy. As the storm enters its main impact phase, with the hurricane approaching within 500 km, wind speeds rise sharply and become highly variable. At the same time, GNSS-IR retrieval errors increase significantly, often exceeding ±50 cm, and become more scattered. After the storm surge, as the hurricane moves away and the wind weakens, the sea surface reflection environment gradually stabilizes. As a result,

GNSS-IR retrieval deviations become smaller again and begin to converge. We conducted a comprehensive assessment of hurricane impacts across 300, 500, and 700 km ranges (Figures S1 and S2). We found that a 500 km radius can encompass and represent the entire storm surge cycle. The 300 km radius captures only the most intense phase of the storm surge, while the 700 km radius includes periods partially unaffected by the storm surge.

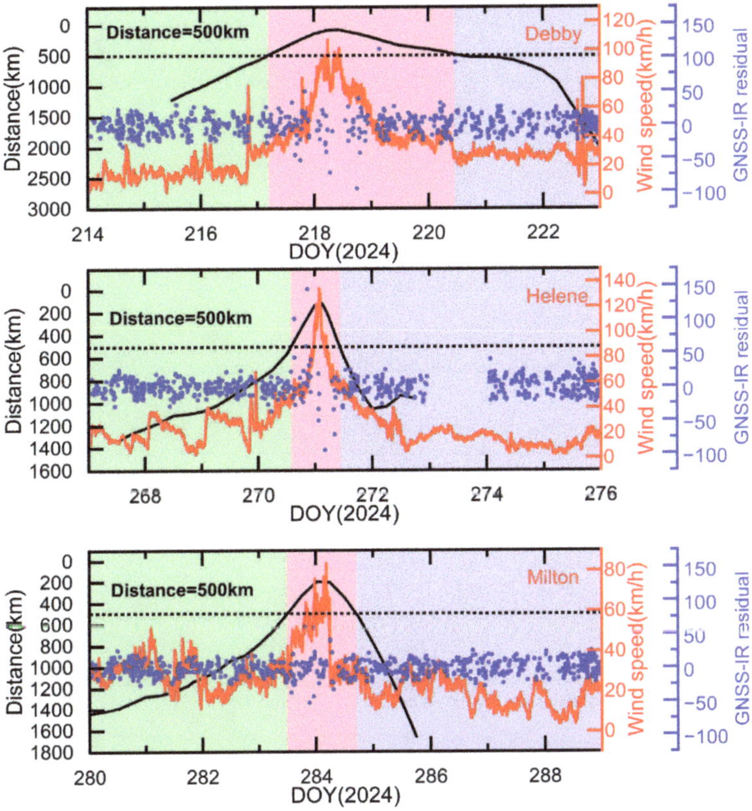

Figure 8. Variations in GNSS-IR retrieval deviations at the FLCK station during three storm surge events in relation to wind speed and the distance between the hurricane eye and the station.

By combining the distance from the eye of the storm and changes in GNSS-IR measurement accuracy, we can identify the specific impact of the spatial movement of such extreme events on local areas. This impact can represent the threat posed by hurricane events to coastal areas, particularly in terms of changes in the storm surge amplitude. We selected 500 km as the boundary for the assessment process, which clearly shows the different degrees of impact on water levels and indicates the deterioration of sea surface reflection/scattering conditions.

The multi-system GNSS retrieval accuracy during each phase of the six storm surge events is summarized in Figures 9 and 10. Bar charts illustrate the RMSE of GNSS across different time periods. The error level during the storm surge phase stands out as substantially higher than in the periods before or after. At the CALC station, the RMSE stays below 9 cm before and after the surge but rises above 15 cm during the storm (Figure 9). At the FLCK station, retrieval errors in the pre- and post-surge periods are generally higher than those at CALC but remain within 13 cm (Figure 10). A particularly large error is observed during Hurricane Helene, where the RMSE reaches 47.55 cm, likely due to a combination of intense wind fluctuations, poor reflection conditions, and fewer visible satellites. We also compiled accuracy statistics for a single system (GPS, Galileo, and Glonass) during storm surge periods and non-storm surge periods (calm period) across six hurricane events (Tables S1–S6). These findings clearly indicate that the GNSS-IR performance is strongly affected by hurricane movement. As the hurricane eye moves closer to the station, increasing wind intensity disrupts the reflection environment, which leads to greater signal degradation and larger retrieval errors. When the hurricane passes and local conditions stabilize, retrieval accuracy improves accordingly. The observed variation highlights the close link between the GNSS-IR error, sea surface condition, and hurricane migration.

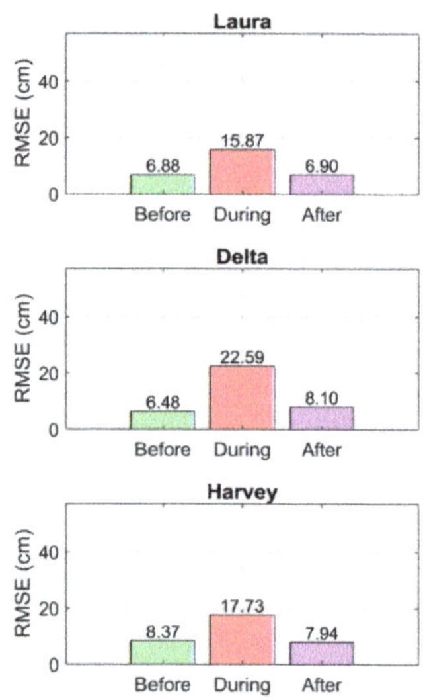

Figure 9. Root mean square error (RMSE) of GNSS-IR-retrieved sea surface heights in three different periods across three storm surge events at the CALC station.

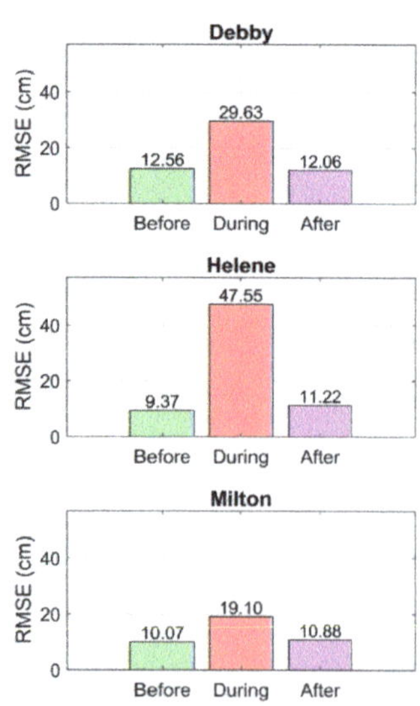

Figure 10. Root mean square error (RMSE) of GNSS-IR-retrieved sea surface heights in three different periods across three storm surge events at the FLCK station.

4.2. Evaluation of Temporal Coverage in GNSS-IR Retrievals

This section presents an analysis of the availability and continuity of GNSS-IR retrievals from multiple systems and frequency bands during the storm surge impact periods. To assess the temporal resolution of GNSS-IR observation sequences, we conducted a statistical analysis using one-hour intervals. For example, within the 0–1 h interval, we determined whether valid observations existed (i.e., the number of observations was greater than or equal to one). If valid observations were present, we concluded that observation coverage existed. For each of the six selected storm surge events, the effective retrieval durations were computed for all frequency bands of GPS, GLONASS, and Galileo over the primary 2-day impact period. Subsequently, time coverage rates were derived for each frequency band to evaluate their applicability and continuity in storm surge monitoring. Figure 11 illustrates the temporal distribution of retrieval results across different GNSS systems and frequency bands for each storm surge event. Overall, the GPS system exhibits higher time coverage than the other systems, benefiting from its larger number of satellites. In particular, the L1 and L2 bands achieve hourly coverage rates ranging between 60% and 80%. The GLONASS system generally attains coverage rates of 50% to 60% across its frequency bands, slightly lower than the GPS. Due to a more limited number of operational satellites, Galileo shows a relatively lower single-frequency coverage, typically between 40% and 50% in most events.

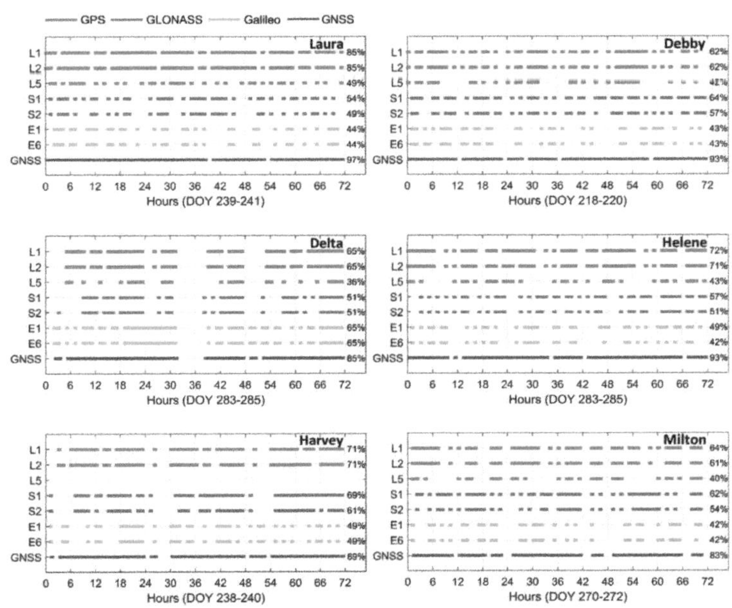

Figure 11. Temporal coverage of GNSS-IR retrievals during storm surge impact periods.

Combining multi-system and multi-frequency retrievals significantly improves temporal coverage. The aggregated data indicate that the overall time coverage during storm surge periods commonly exceeds 85%, with Hurricane Laura reaching as high as 97%. These results suggest that GNSS-IR technology, when integrating data from multiple GNSS systems and frequency bands, can provide near-continuous monitoring of storm surge processes on an hourly timescale. The integration of multiple GNSS systems not only significantly increases the spatial and temporal density of water level observations but also enhances the robustness and practicality of the GNSS-IR approach under extreme weather conditions.

4.3. Analysis of Factors Affecting GNSS-IR Monitoring Accuracy

Table 3 reveals a significant anomaly in the retrieval accuracy of the GPS L2 frequency band at the FLCK station, particularly during Hurricane Milton, where the RMSE exceeds 90 cm—far above the typical error range. To investigate the source of this error, a detailed analysis of the Lomb–Scargle Periodogram (LSP) results for the L2 band was conducted. Taking Hurricane Helene as an example, Figure 12 presents the retrieved reflector heights (RH, in meters) and corresponding spectral analysis results for the G08 satellite during DOY 280–288, with the L1 band shown as a gray curve and the L2 band as a purple curve. The figure shows that the L2 band exhibits multiple peaks exceeding two in several time intervals, with incorrect peaks often dominating as the primary peak, leading to a pronounced overestimation. For a systematic investigation, the difference between the primary frequency retrieval heights of the L2 and L1 bands at FLCK was calculated. The results showed that the L2-derived heights were generally lower than those from L1, with most differences concentrated around −2.70 m, indicating the presence of a systematic bias in the L2-band retrievals. Considering the actual environment of the FLCK station, the surrounding area includes not only the sea surface but also hard structures such as piers, resulting in a complex reflecting surface. In this specific case, G08, which passes through a specific space quadrant every consecutive day, as a result, produces similar outstanding reflected signals during particular minute spans of the day (this is clearly shown by Figure 12). This setting may increase the likelihood and intensity of the "double peak phenomenon" in the L2 band [32]. Additionally, sea surface conditions become more complex and unstable during storm surges, leading to a significantly degraded reflection quality. These combined factors may contribute to the reduced accuracy of L2-band retrievals.

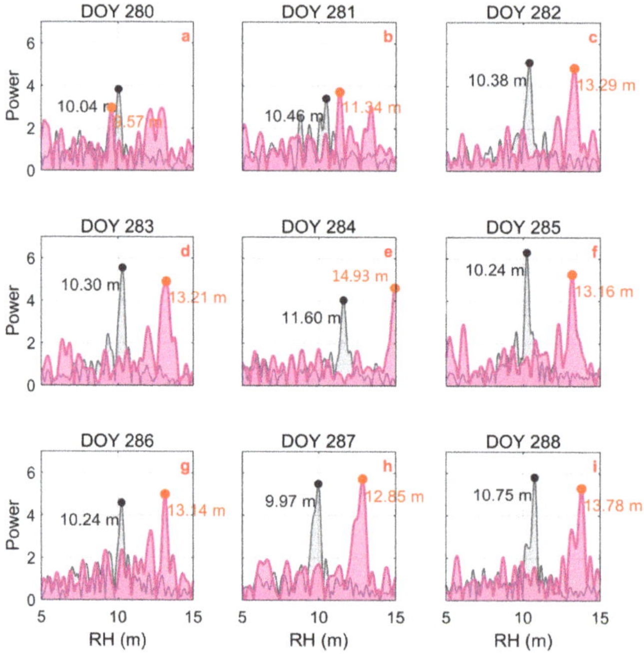

Figure 12. Comparison of Lomb–Scargle Periodogram (LSP) results for GPS L2 and L1 bands at FLCK station during DOY 280–288 in 2024, using satellite G08 as an example, from Figure (**a**–**i**) is the LSP result from DOY280 to DOY288. The dominant frequencies for each band are labeled to identify the corresponding reflection heights (RH) of the sea surface reflected signals.

5. Conclusions

This study focused on two typical coastal GNSS stations, CALC and FLCK, and analyzed GNSS-IR sea surface height retrievals during six representative storm surge events in the Gulf of Mexico. The applicability of GNSS-IR technology under storm surge conditions was validated, with an in-depth discussion on retrieval accuracy, spatiotemporal coverage, and site environment adaptability. The main conclusions are as follows:

1. GNSS-IR technology enables the effective monitoring of sea surface height during storm surge events. By integrating data from multiple GNSS systems and frequency bands, the spatiotemporal resolution and overall accuracy of the retrievals can be significantly improved. Specifically, the time coverage rate exceeds 83% with multi-GNSS integration. In calm conditions, the root mean square error (RMSE) of sea level retrievals can reach ~7 cm. During storm surge periods, however, the accuracy degrades markedly to the decimeter level, and can exceed 30 cm under peak surge conditions, especially when the hurricane eye is close. In addition, the site environment (water-reflected condition) is a key factor, which dominates the worse precision of the FLCK site.
2. Storm surge processes have a pronounced impact on GNSS-IR retrieval accuracy, particularly when the hurricane eye approaches or passes near the station. The decrease in accuracy closely correlates with the hurricane's migration path and sea surface reflection/scattering conditions.
3. GNSS-IR technology demonstrates high temporal coverage in tide level monitoring. The temporal coverage rate is defined as the percentage of 1 h intervals containing at least one valid sea level retrieval. For a single satellite navigation system, this coverage typically ranges from 40% to 60%. By integrating multiple systems, the overall coverage can be increased to over 85%, as shown in Figure 11, which demonstrates that GNSS-IR has the capability to conduct the continuous monitoring of daily tidal variations and observe extreme events such as storm surges.
4. At the FLCK station, the GPS L2 frequency band exhibits a generally poor retrieval accuracy with systematic bias, likely related to the complex non-water reflective environment around the site and the L2 signal's susceptibility to interference.

The application of GNSS-IR technology is becoming increasingly widespread, covering multiple scenarios such as rivers, oceans, lakes, and reservoirs. Our research has demonstrated the feasibility and reliability of GNSS-IR technology under extreme weather conditions. In the future, GNSS can make more contributions to hydrological and ocean monitoring, providing data support and technical services for global climate change monitoring and research.

Supplementary Materials: The following supporting information can be downloaded at: https://www.mdpi.com/article/10.3390/rs17183132/s1. Figure S1: Relationship between GNSS-IR retrieval bias and the distance between the hurricane eye and the CALC station, evaluated within three distance thresholds: 300 km, 500 km, and 700 km. Dashed lines indicate ±10 cm, ±20 cm, and ±30 cm error margins. Figure S2: Relationship between GNSS-IR retrieval bias and the distance between the hurricane eye and the FLCK station, evaluated within three distance thresholds: 300 km, 500 km, and 700 km. Dashed lines indicate ±10 cm, ±20 cm, and ±30 cm error margins. Table S1: Performance Metrics of GNSS-IR for Hurricane Laura: Number of Observations (Num), Root Mean Square Error (RMSE, cm), Systematic Error (Bias, cm), Mean Absolute Error (MAE, cm), Pearson Correlation Coefficient (R), and Nash–Sutcliffe Efficiency (NSE). Table S2: Performance Metrics of GNSS-IR for Hurricane Delta: Num, RMSE (cm), Bias (cm), MAE (cm), R, and NSE under Calm and Storm Conditions (same metrics as Table S1). Table S3: Performance Metrics of GNSS-IR for Hurricane Harvey: Num, RMSE (cm), Bias (cm), MAE (cm), R, and NSE under Calm and Storm

Conditions (same metrics as Table S1). Table S4: Performance Metrics of GNSS-IR for Hurricane Debby: Num, RMSE (cm), Bias (cm), MAE (cm), R, and NSE under Calm and Storm Conditions (same metrics as Table S1). Table S5: Performance Metrics of GNSS-IR for Hurricane Helene: Num, RMSE (cm), Bias (cm), MAE (cm), R, and NSE under Calm and Storm Conditions (same metrics as Table S1). Table S6: Performance Metrics of GNSS-IR for Hurricane Milton: Num, RMSE (cm), Bias (cm), MAE (cm), R, and NSE under Calm and Storm Conditions (same metrics as Table S1).

Author Contributions: Conceptualization and methodology, K.L.; software and validation, R.Z.; resources and data curation, X.W. and T.X.; writing—original draft preparation, R.Z. and K.L.; writing—review and editing, Q.C. and Z.L.; visualization, R.Z. and X.C. All authors have read and agreed to the published version of the manuscript.

Funding: This research is funded by the Natural Science Foundation of Wuhan [grant No. 2024040701010029], the Postdoctor Project of Hubei Province [grant No. 2024HBBHCXB060], the Open Fund of Technology Innovation Center for Geohazard Monitoring and Risk Early Warning, Ministry of Natural Resources [grant No. TICGM-2024-01], and the Basic Science Center Project of the National Natural Science Foundation of China [grant No. 42388102]. This research also is supported by National Key Research and Development Program of China [grant No. 2024YFC3810502] and State Key Laboratory of Geodesy and Earth's Dynamics, Innovation Academy for Precision Measurement Science and Technology, Chinese Academy of Sciences [grant No. SKLPG2025-5-1].

Data Availability Statement: The GPS GLONASS and Galileo observations are available at ftp://igs.gnsswhu.cn/pub/gps/data/daily/ (accessed on 26 August 2025). The water level at CALC and FLCK are available at https://tidesandcurrents.noaa.gov/waterlevels.html (accessed on 26 August 2025). The wind data at CALC and FLCK are available at https://tidesandcurrents.noaa.gov/met.html (accessed on 26 August 2025). The locations of eyes of hurricanes are available at https://www.nhc.noaa.gov/data/hurdat/hurdat2-1851-2024-040225.txt (accessed on 26 August 2025).

Acknowledgments: We thank the NOAA for providing water level, wind speed, and locations of eyes of hurricanes that we used in this research.

Conflicts of Interest: Author Xue Wang was employed by the company Beijing FreeDo Technology Co., Ltd. The remaining authors declare that the research was conducted in the absence of any commercial or financial relationships that could be construed as a potential conflict of interest.

References

1. Muis, S.; Verlaan, M.; Winsemius, H.; Aerts, J.C.J.H.; Ward, P.J. A global reanalysis of storm surges and extreme sea levels. *Nat. Commun.* **2016**, *7*, 11969. [CrossRef] [PubMed]
2. Gabe, T.; Falk, G.; McCarty, M.; Mason, V.W. Hurricane katrina: Social-demographic characteristics of impacted areas. In *CRS Report for Congress*; CRS: Boca Raton, FL, USA, 2005; Available online: https://sgp.fas.org/crs/misc/RL33141.pdf (accessed on 4 September 2025).
3. Eley, E.N.; Subrahmanyam, B.; Trott, C.B. Ocean–Atmosphere Interactions during Hurricanes Marco and Laura. *Remote Sens.* **2021**, *13*, 1932. [CrossRef]
4. Nicholls, R.J.; Lincke, D.; Hinkel, J.; Brown, S.; Vafeidis, A.T.; Meyssignac, B.; Hanson, S.E.; Merkens, J.-L.; Fang, J. A global analysis of subsidence, relative sea-level change and coastal flood exposure. *Nat. Clim. Change* **2021**, *11*, 338–342. [CrossRef]
5. Wang, J.; Gao, W.; Xu, S.; Yu, L. Evaluation of the combined risk of sea level rise, land subsidence, and storm surges on the coastal areas of Shanghai, China. *Clim. Change* **2012**, *115*, 537–558. [CrossRef]
6. Ablain, M.; Cazenave, A.; Larnicol, G.; Balmaseda, M.; Cipollini, P.; Faugère, Y.; Fernandes, M.J.; Henry, O.; Johannessen, J.A.; Knudsen, P.; et al. Improved sea level record over the satellite altimetry era (1993–2010) from the climate change initiative project. *Ocean. Sci.* **2015**, *11*, 67–82. [CrossRef]
7. Lagmay, A.M.F.; Agaton, R.P.; Bahala, M.A.C.; Briones, J.B.L.T.; Cabacaba, K.M.C.; Caro, C.V.C.; Dasallas, L.L.; Gonzalo, L.A.L.; Ladiero, C.N.; Lapidez, J.P.; et al. Devastating storm surges of typhoon haiyan. *Int. J. Disaster Risk Reduct.* **2015**, *11*, 1–12. [CrossRef]
8. Zhang, G.; Kuang, C.; Chen, C. Research on GNSS-R Storm Surge Monitoring Based on Non-Linear Fitting Method. *J. Geod. Geodyn.* **2025**, *45*, 500–505, 531. [CrossRef]
9. Vignudelli, S.; Birol, F.; Benveniste, J.; Fu, L.-L.; Picot, N.; Raynal, M.; Roinard, H. Satellite Altimetry Measurements of Sea Level in the Coastal Zone. *Surv. Geophys.* **2019**, *40*, 1319–1349. [CrossRef]

10. Birol, F.; Fuller, N.; Lyard, F.; Cancet, M.; Niño, F.; Delebecque, C.; Fleury, S.; Toublanc, F.; Melet, A.; Saraceno, M.; et al. Coastal applications from nadir altimetry: Example of the X-TRACK regional products. *Adv. Space Res.* **2017**, *59*, 936–953. [CrossRef]
11. Liu, K.; Wen, Y.; Zeng, J.; Li, Z.; Xu, C. Rapid early afterslip characteristics of the 2010 moment magnitude (Mw) 8.8 Maule earthquake determined with sub-daily GPS solutions. *Satell. Navig.* **2024**, *5*, 23. [CrossRef]
12. Larson, K.M.; Small, E.E.; Gutmann, E.; Bilich, A.; Axelrad, P.; Braun, J. Using GPS multipath to measure soil moisture fluctuations: Initial results. *GPS Solut.* **2008**, *12*, 173–177. [CrossRef]
13. Larson, K.M.; Löfgren, J.S.; Haas, R. Coastal Sea Level Measurements Using a Single Geodetic GPS Receiver. *Adv. Space Res.* **2013**, *51*, 1301–1310. [CrossRef]
14. Löfgren, J.S.; Haas, R.; Scherneck, H.-G. Sea level time series and ocean tide analysis from multipath signals at five GPS sites in different parts of the world. *J. Geodyn.* **2014**, *80*, 66–80. [CrossRef]
15. Chen, L.; Chai, H.; Zheng, N.; Wang, M.; Xiang, M. Feasibility and performance evaluation of low-cost GNSS devices for sea level measurement based on GNSS-IR. *Adv. Space Res.* **2023**, *72*, 4651–4662. [CrossRef]
16. Cahyadi, M.; Bawasir, A.; Susilo; Arief, S. Analysis of Water Level Monitoring using GNSS Interferometric Reflectometry in River Waters. *IOP Conf. Ser. Earth Environ. Sci.* **2023**, *1276*, 012020. [CrossRef]
17. Holden, L.D.; Larson, K.M. Ten years of Lake Taupō surface height estimates using the GNSS interferometric reflectometry. *J. Geod.* **2021**, *95*, 74. [CrossRef]
18. Wang, P.; Tu, R.; Wang, X.; Han, J.; Zhang, J.; Cheng, F.; Lu, X. GNSS-IR water level monitoring for complex environments: Application to Kalabeily Reservoir in Xinjiang, China. *Adv. Space Res.* **2025**, *75*, 7035–7048. [CrossRef]
19. Wang, X.; He, X.; Xiao, R.; Song, M.; Jia, D. Millimeter to centimeter scale precision water-level monitoring using GNSS reflectometry: Application to the south-to-north water diversion project, China. *Remote Sens. Environ.* **2021**, *265*, 112645. [CrossRef]
20. Shan, Q.; Chen, Q.; Liu, Y.; Jiang, W.; Liu, K.; Zhou, X. The potential of GNSS-IR for monitoring daily water level fluctuations up to 15 m. *Measurement* **2025**, *249*, 117053. [CrossRef]
21. Wang, X.; Zhang, Q.; Zhang, S. Sea level estimation from SNR data of geodetic receivers using wavelet analysis. *GPS Solut.* **2019**, *23*, 6. [CrossRef]
22. Zhang, S.; Liu, K.; Liu, Q.; Zhang, C.; Zhang, Q.; Nan, Y. Tide variation monitoring based improved GNSS-MR by empirical mode decomposition. *Adv. Space Res.* **2019**, *63*, 3333–3345. [CrossRef]
23. Peng, D.; Hill, E.M.; Li, L.; Switzer, A.D.; Larson, K.M. Application of GNSS interferometric reflectometry for detecting storm surges. *GPS Solut.* **2019**, *23*, 47. [CrossRef]
24. Vu, P.L.; Ha, M.C.; Frappart, F.; Darrozes, J.; Ramillien, G.; Dufrechou, G.; Gegout, P.; Morichon, D.; Bonneton, P. Identifying 2010 xynthia storm signature in GNSS-R-based tide records. *Remote Sens.* **2019**, *11*, 782. [CrossRef]
25. Wang, X.; He, X.; Shi, J.; Chen, S.; Niu, Z. Estimating sea level, wind direction, significant wave height, and wave peak period using a geodetic GNSS receiver. *Remote Sens. Environ.* **2022**, *279*, 113135. [CrossRef]
26. Zhang, T.; Wang, X. Tide inversion based on GNSS-IR and analysis of barometric correlation under meteorological forcing. *GNSS World China* **2025**, *50*, 70–78. [CrossRef]
27. Larson, K.M.; Lay, T.; Yamazaki, Y.; Cheung, K.F.; Ye, L.; Williams, S.D.P.; Davis, J.L. Dynamic sea level variation from GNSS: 2020 Shumagin earthquake tsunami resonance and Hurricane Laura. *Geophys. Res. Lett.* **2021**, *48*, e2020GL091378. [CrossRef]
28. Chai, H.; Chen, K. Facilitated interferometric reflectometry evaluation and its application in monitoring three typhoon storm surges in hong kong with multi-GNSS constellation. *GPS Solut.* **2024**, *28*, 99. [CrossRef]
29. Li, L.; Qiu, Q.; Ye, M.; Peng, D.; Hsu, Y.-J.; Wang, P.; Shi, H.; Larson, K.M.; Zhang, P. Island-based GNSS-IR network for tsunami detecting and warning. *Coast. Eng.* **2024**, *190*, 104501. [CrossRef]
30. Bilich, A.; Larson, K.M. Mapping the GPS multipath environment using the signal-to-noise ratio (SNR). *Radio Sci.* **2007**, *42*, RS6003. [CrossRef]
31. Feng, P.; Haas, R.; Elgered, G. A novel tropospheric error formula for ground-based GNSS interferometric reflectometry. *IEEE Trans. Geosci. Remote Sens.* **2023**, *61*, 5802918. [CrossRef]
32. Wang, X.; Song, M.; He, X. Simulation and Correction of Double-Peak Errors in GPS L2P(Y) Signals for GNSS-IR Applications. *IEEE Geosci. Remote Sens. Lett.* **2025**, *22*, 3001205. [CrossRef]

remote sensing

MDPI

Article

Estimation of Downlink Signal Transmitting Antenna PCO and Equipment Delays for LEO Navigation Constellations with Limited Regional Stations

Ziqiang Li, Wanke Liu * and Jie Hu

School of Geodesy and Geomatics, Wuhan University, Wuhan 430079, China; ziqiangli@whu.edu.cn (Z.L.); jiehu1995@whu.edu.cn (J.H.)

* Correspondence: wkliu@sgg.whu.edu.cn

Highlights:

What are the main findings?

- A joint estimation framework is proposed to estimate the downlink signal PCO and equipment delay for LEO constellations using limited regional stations.
- Extensive simulations under varying orbit altitudes, constellation scales, and error configurations validate the effectiveness and robustness of the proposed method.

What is the implication of the main finding?

- The proposed method enhances the practicality of PCO and delay calibration for real-world LEO navigation augmentation systems.
- It provides an effective strategy for error modeling and system design under sparse tracking conditions.

Abstract: In LEO constellation–augmented navigation, the transmitting antenna phase center offset (PCO) and the equipment delay associated with the downlink signals of LEO satellites constitute major error sources that must be precisely characterized. Previous studies primarily focused on single or small-scale satellite scenarios, lacking comprehensive evaluations regarding the influence of constellation scale, orbital altitude, ground station configuration, and various error sources. To address this gap, we propose a joint estimation method utilizing observations from a limited number of regional ground stations in China that simultaneously track GNSS and LEO satellites. The method is specifically designed to accommodate practical constraints on ground station distribution within China. Initially, a batch least-squares estimation strategy is employed to simultaneously determine the ionosphere-free PCO and initial equipment delays for all LEO satellites in a constellation-wide solution. Subsequently, the estimated PCO parameters are fixed, and the equipment delays are further refined using a precise point positioning (PPP) approach. To systematically evaluate the method's performance under realistic conditions, we analyze the impact of orbital altitude, constellation size, ground station number, data processing duration, and orbit/clock biases through comprehensive simulations. The results indicate: (1) the Z-direction component of the PCO (pointing toward the Earth's center) and equipment delay is more sensitive to orbit and clock errors; (2) Increasing the number of LEO satellites generally improves the estimation accuracy of equipment delays, but the marginal gain diminishes as the constellation size expands; (3) sub-centimeter PCO accuracy and equipment delay accuracies better than 3 cm can still be achieved using only 3–4 regionally distributed ground stations over an observation period of 5–7 days.

Keywords: LEO constellation-augmented Navigation; phase-center offset (PCO); equipment delay

1. Introduction

As a technological revolution after RTK and network RTK technology, Precise point positioning (PPP), with advantages such as stand-alone operation and flexible mobility, is widely used in location service [1–5]. Nevertheless, GNSS PPP typically requires tens of minutes for convergence without a precise atmospheric parameter, presenting a significant hurdle to the real-time applications. Benefiting from the rapid geometric changes, Low Earth Orbit (LEO) satellites can expedite the convergence process of GNSS PPP [6,7]. Thus, LEO-augmented PPP has attracted considerable attention from researchers, which is poised to be a pivotal component of the next-generation satellite navigation system [8–11]. The basic mathematical model of LEO-augmented PPP is established based on the conventional GNSS observation model. In addition, the contribution of the LEO navigation enhancement constellation to PPP has been systematically analyzed in previous studies [12–14]. Ke et al. (2015) simulated and compared the positioning accuracy and convergence time of GPS-only PPP, GPS/GLONASS PPP and GPS/LEO PPP, which shows that LEO satellites have more contribution to PPP convergence than MEO satellites [15]. Ge et al. (2024) analyzed the impact of LEO satellites on the spatial geometry based on simulated LEO and real GNSS data, and evaluated the LEO constellation-augmented Navigation performance with ionosphere-free (IF), undifferenced and uncombined (UDUC), and ionosphere-weighted (IW) models [16]. Li et al. (2019) comprehensively analyzed the influence of constellation configuration with simulation experiments [17]. Li et al. (2023) and Chang et al. (2025) analyzed the impact of LEO on PPP in terms of convergence time and positioning accuracy, integrating CENTISPACET™ satellites with GNSS [18,19]. The above studies demonstrate the superiority of LEO-augmented PPP in convergence.

Despite the similarities in mathematic models between LEO-augmented and traditional PPP, the unique challenges introduced by the deployment of new LEO constellations, which have also become studies' focus beyond convergence performance. The technology and experience of utilizing space-borne GPS data for post-processing precise orbit determination of satellites, both domestically and internationally, have reached a high level of maturity. Satellite series such as GOCE, CHAMP, GRACE, HY2A, and the JASON series satellites, among others, are all capable of achieving centimeter-level orbit accuracy [20–24]. The LEO satellite orbit calculated by on-board GNSS data is referenced to the center of mass, and the clock error is relative to the LEO onboard GNSS receiver [25–28]. To unify the spatial relationship in LEO-augmented PPP, it is necessary to convert the center-of-mass orbit to the phase-center references of the downlink navigation signals. Hence, the high-precision phase-center offsets (PCO) of the downlink navigation signals with respect to the center-of-mass must be determined to convert the orbit parameter. Typically, there is an initial calibration value for the downlink signal PCO on the ground. The clock error relative to the onboard GNSS receiver must also be converted to the downlink signal transmitter, which we call the downlink signal transmitter equipment delay relative to the onboard GNSS receiver here. Both types of hardware delays can be calibrated on the ground before launching LEO satellites. For example, Li et al. (2023) adopted factory-calibrated downlink antenna PCO values for correction in the CENTISPACET-based LEO augmentation experiments [18]. However, during in-orbit operation, factors such as structural deformation and shifts in the satellite's center of mass can cause the actual phase center to deviate from the pre-launch calibration. Such deviations may reach several centimeters, and equipment delays may also drift over time, making it impractical to rely solely on static calibration values for long-term applications [29]. Moreover, due to the low orbital altitude, rapid geometric variation, and short ground visibility of LEO satellites, along with the foreseeable limitation in the number of ground stations capable of receiving LEO navigation signals within China, it is essential to develop dedicated algorithms for estimating the downlink signal PCO and equipment delay.

The equipment delay, as an inherent component of the pseudorange and carrier-phase measurements from downlink signals received at ground monitoring stations, exhibits different physical characteristics from the downlink signal transmitting antenna phase center offset (PCO). While the PCO is generally a stable geometric parameter, the equipment delay is more susceptible to temperature variations and electronic fluctuations, showing time-varying behavior. Therefore, these two parameters should be estimated simultaneously using different modeling strategies.

In this context, GNSS observations can provide high-precision solutions for station-related parameters such as receiver clock offset and ZTD, serving as important auxiliary information for the estimation of LEO downlink parameters [18,30]. However, when combining GNSS and LEO observations, inter-system biases (ISB) between the two systems may arise and must be properly accounted for. Liu et al. (2024) [30] explored the feasibility of jointly estimating the downlink signal PCO and equipment delay of LEO satellites using pseudorange and carrier-phase observations from GNSS and LEO downlinks. In their simulations based on Sentinel-3B and Sentinel-6A downlink data over 10 globally distributed stations for 6 processing days, the post-processed least-squares solution equipment delay 3.5 mm for PCO and 3 cm for equipment delay. Additionally, Li et al. (2023) [18] proposed a "Three-Step Approach" in their experiments with CENTISPACET satellites. This method first uses GNSS data to estimate station clock offsets and tropospheric delays, then fixes these parameters to estimate LEO satellite equipment delays and ambiguities via static PPP, and finally derives epoch-wise delay corrections. This approach is particularly effective under conditions with only a small number of LEO satellites.

Although Liu et al. (2024) and Li et al. (2023) have made valuable contributions to the estimation of downlink signal PCO and equipment delay for LEO satellites, their studies are primarily based on single or a small number of experimental satellites under static conditions [18,30]. For practical deployment in future large-scale LEO navigation constellations, several key limitations remain:

(1) Existing approaches typically assume that receiver clock offsets and tropospheric delays at ground stations are already estimated using GNSS data, and that LEO observations serve solely to estimate satellite-side parameters. The potential of large volumes of LEO observations to enhance the estimation of ground station parameters has not been explored, particularly under constellation-scale expansion.
(2) There is currently no comprehensive evaluation of how key system factors—such as LEO satellite orbital altitude, constellation size, and orbit/clock bias error levels—affect the accuracy of PCO and equipment delay estimation. This limits support for constellation design and performance optimization.
(3) Given the practical limitations of ground station deployment in China, it remains unclear whether reliable estimation of LEO downlink signal PCO and equipment delays is achievable under conditions with only a small number of regionally ground stations within China.

To address the aforementioned issues, this study proposes a joint estimation method for the downlink signal PCO and equipment delay of LEO satellites. The method builds upon pseudorange and carrier-phase observations collected from ground stations simultaneously tracking both LEO and GNSS downlink signals. In the first stage, a batch least-squares strategy is employed to estimate the parameters of the PCO and posteriori equipment delay. With the estimated PCO parameters fixed, the equipment delays are further refined using a PPP approach. Notably, the method is designed with practical constraints in mind, particularly the limited deployment of ground stations in China. Therefore, the estimation framework is implemented and validated using only a limited number of regionally distributed ground stations within China. Additionally, this study conducts a systematic evaluation of how factors such as orbital altitude, constellation size, number of ground stations, and orbit/clock bias errors affect the estimation performance.

2. Downlink Signals Transmitting Antenna PCO and Equipment Delay Estimation of LEO Satellites

As mentioned before, the LEO satellite orbit and clock error are estimated with respect to the center-of-mass rather than the phase-center references of the downlink navigation signals, which is essential to consider the PCO and equipment delay of downlink signals for LEO satellites. Figure 1 shows spacecraft body-fixed system and different antenna systems of the LEO satellite. To ensure the reliable estimation of these parameters, it is necessary to incorporate accurate station positions, station clock bias, and tropospheric delay corrections. GNSS observations provide high-precision solutions for these related parameters, which can serve as critical auxiliary information for estimating the downlink parameters of LEO satellites. Thus, it is necessary to jointly process the GNSS satellite observations and LEO downlink navigation observations.

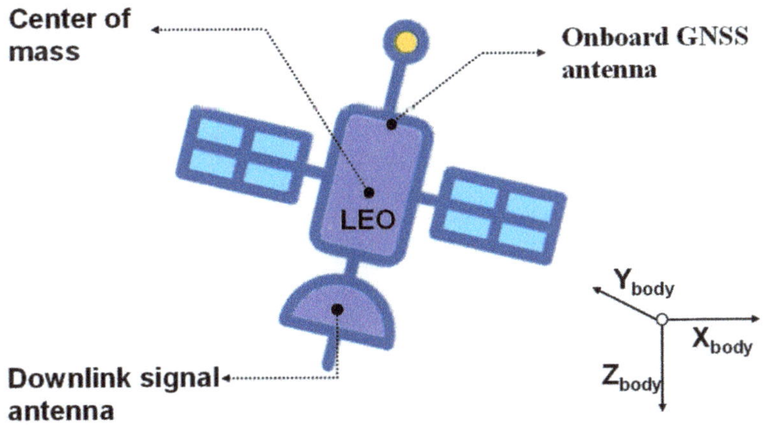

Figure 1. Spacecraft body-fixed system and different antenna systems of the LEO satellite.

2.1. Mathematical Model

The LEO satellite downlink signal PCO and equipment delay are included in the pseudorange and phase observations of the ground receivers. The impact of PCO errors on downlink observations differs from that of equipment delays due to their distinct physical characteristics [30]. Thus, the two parameters can be separated due to their distinct characteristics in observations. Since the ionosphere-free combination is commonly used in the LEO-augmented PPP, we mainly focus on the dual-frequency ionosphere-free equivalent values of PCO and equipment delay for LEO satellites.

The pseudorange and phase observation include satellite-related errors, receiver-related errors, and signal transmission-related errors. Since the relativistic effects, earth rotation effect and tide effect can be precisely corrected with model. The observation equation can be simplified as follows [31–34]:

$$\begin{cases} P_r^{s,k} = \rho_r^{s,k} - t^{s,k} + t_r^k + I_r^k + T_r^k + e_r^{s,k} \\ \phi_r^{s,k} = \rho_r^{s,k} - t^{s,k} + t_r^k - I_r^k + T_r^k + \lambda_{s,k} N_r^{s,k} + \varepsilon_r^{s,k} \end{cases} \quad (1)$$

where s and r denote the satellite and the ground receiver respectively; k stands for the satellite system (BDS or LEO); λ is the satellite wavelength; ρ is the geometric distance between the satellite antenna phase center and the receiver antenna phase center; t is the clock error; I is the Ionospheric delay; T is the tropospheric delay. e denotes the measurement noise and multipath in the satellite pseudorange observations, and ε denotes the measurement noise and multipath in the carrier-phase observations, respectively.

The ionospheric-free combination can eliminate the effect of first-order ionospheric delay. Thus, the ionospheric-free combination model of the BDS satellite and the LEO satellite can be derived as follows:

$$\begin{cases} P_{r,IF}^{s,k} = \alpha_{s,k} P_{r,1}^{s,k} + \beta_{s,k} P_{r,2}^{s,k} = \rho_{r,IF}^{s,k} - t^{s,k} + t_r^k + T_r^k + e_{r,IF}^{s,k} \\ \phi_{r,IF}^{s,k} = \alpha_{s,k} \phi_{r,1}^{s,k} + \beta_{s,k} \phi_{r,2}^{s,k} = \rho_{r,IF}^{s,k} - t^{s,k} + t_r^{s,k} + T_r^k + \lambda_{IF}^{s,k} N_{r,IF}^{s,k} + \varepsilon_{r,IF}^{s,k} \end{cases} \quad (2)$$

with:

$$\begin{cases} \alpha_{s,k} = f_{1,k}^2 / \left(f_{1,k}^2 - f_{2,k}^2\right) \\ \beta_{s,k} = -f_{2,k}^2 / \left(f_{1,k}^2 - f_{2,k}^2\right) \end{cases} \quad (3)$$

where IF represents ionospheric-free combination; α and β are the ionospheric-free combination coefficient. $f_{1,k}^2$ and $f_{2,k}^2$ represent the signal frequencies used by the GNSS satellite and the LEO satellite, respectively.

The geometric distance $\rho_{r,IF}^{s,k}$ between the satellite downlink signal antenna phase center and the receiver phase center is calculated with the center-of-mass position for the satellite and receiver according to the PCO correction. The procedure can be depicted with the following equation [34,35]:

$$\rho_{r,IF}^{s,k} = \left| (\boldsymbol{p_r} + \boldsymbol{pco_{r,IF}}) - (\boldsymbol{p_{IF}^{s,k}} + \boldsymbol{pco_{IF}^{s,k}}) \right| \quad (4)$$

with

$$\boldsymbol{pco_{IF}^{s,k}} = \boldsymbol{e_{IF}^{s,k}} \cdot \boldsymbol{pco'}_{\boldsymbol{IF}}^{\boldsymbol{s,k}} \quad (5)$$

where p_r denotes the center-of-mass position vector of the ground antenna; $\boldsymbol{pco_{r,IF}}$ denotes the PCO vector of the ground receiver, defined from the receiver's center of mass to the antenna phase center; $\boldsymbol{p_{IF}^{s,k}}$ denote the center-of-mass position vector for the satellite; $\boldsymbol{pco_{IF}^{s,k}}$ denotes the PCO vector of the satellite in the earth-fixed system; $\boldsymbol{pco'}_{\boldsymbol{IF}}^{\boldsymbol{s,k}}$ denotes the satellite PCO vector in the satellite-fixed coordinate systems, which can be shown in Figure 2; $\boldsymbol{e_{IF}^{s,k}}$ is the rotation matrix from the spacecraft body-fixed system to the earth-fixed system, which is determined by the satellite attitude control mode. The attitude model of BDS satellites follows the definition provided in the BeiDou Signal-in-Space Interface Control Documentd [35]. All LEO satellites used in this study are assumed to be three-axis stabilized, which is consistent with the configuration of most LEO satellites deployed for navigation or Earth observation missions.

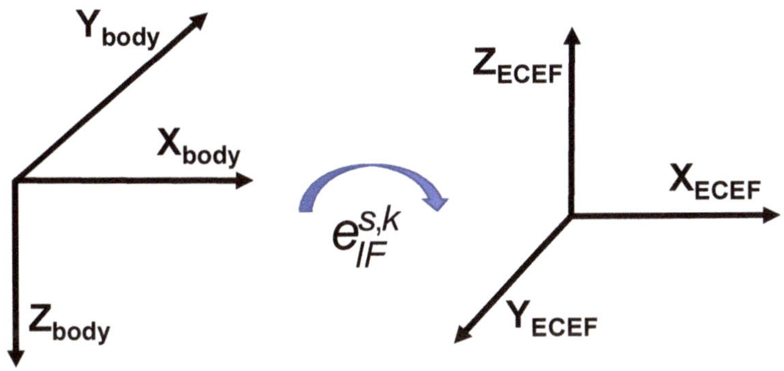

Figure 2. Conversion of satellite-fixed coordinate system to earth-fixed coordinate system.

In the LEO-augmented PPP, the clock error relative to the onboard GNSS receiver must be converted to downlink signal transmitter. An equipment delay introduces the observation equation:

$$t^{s,L} = b_{IF}^{s,L} + \bar{t}^{s,L} \tag{6}$$

Under the condition of a sparse regional ground station network, where station numbers are limited and spatial geometry is weak, the observation model must be carefully adapted to ensure separability between satellite-side and ground-side parameters. An inter-system bias (ISB) term is introduced for each station to absorb residual differences between the GNSS and LEO systems. The combined use of LEO and GNSS observations enables joint estimation of receiver clock offsets, tropospheric delays, LEO satellite PCOs, and equipment delays. The systematic bias between LEO and GNSS time system is also introduced in the observation equation:

$$t_r^L = t_r^G + ISB_L \tag{7}$$

Combining Equations (2)–(7), the unified observation model for estimating the downlink signal PCO and equipment delays of the full LEO constellation can be expressed as follows:

$$\begin{bmatrix} v_{P,r}^L \\ v_{\phi,r}^L \\ v_{P,r}^G \\ v_{\phi,r}^G \end{bmatrix} = \begin{bmatrix} \boldsymbol{e}_{\boldsymbol{IF}}^{\boldsymbol{s,L}} & 1 & 1 & 1 & 1 & h^{s,L} & 0 \\ \boldsymbol{e}_{\boldsymbol{IF}}^{\boldsymbol{s,L}} & 0 & 1 & 1 & 1 & h^{s,L} & 0 \\ \boldsymbol{0}_{1*3} & 0 & 0 & 1 & 0 & h^{s,G} & 0 \\ \boldsymbol{0}_{1*3} & 0 & 0 & 1 & 0 & h^{s,G} & 1 \end{bmatrix} \begin{bmatrix} \delta \boldsymbol{pco}_{\boldsymbol{IF}}^{\prime \boldsymbol{s,L}} \\ b_{IF}^{s,L} \\ \delta N_{r,IF}^{s,L} \\ ISB_L \\ \delta t_r^G \\ \delta T_r \\ \delta N_{r,IF}^{s,G} \end{bmatrix} + \begin{bmatrix} l_{P,r}^L \\ l_{\phi,r}^L \\ l_{P,r}^G \\ l_{\phi,r}^G \end{bmatrix} \tag{8}$$

where h denotes the zenith projection of satellite and receiver.

According to Equation (8), LEO downlink signals the PCO and equipment delay can be estimated through jointly processing the GNSS satellite observations and LEO downlink navigation observations.

It is important to note that, in the case of single-satellite estimation, the equipment delay is strongly correlated with ISB_L. As a result, the estimated equipment delay $\delta \bar{b}_{IF}^{s,L}$ in a single-satellite solution inherently includes the ISB_L component, and can be expressed as

$$\bar{b}_{IF}^{s,L} = b_{IF}^{s,L} + ISB_L \tag{9}$$

2.2. Calculation Process

Based on the above analysis, we establish a joint estimation framework for the downlink antenna phase center offset (PCO) and equipment delay of LEO constellation satellites. The overall workflow, as illustrated in Figure 3, consists of the following three main steps:

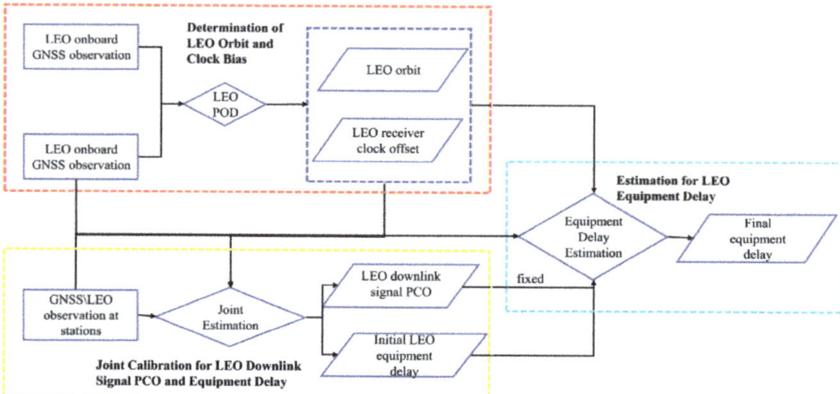

Figure 3. Flowchart of the Calibration Process for LEO Satellite Downlink Signal PCOs and Equipment Delay under a Regional Ground Station Network.

Step 1: The center-of-mass orbits of LEO satellites and the satellite clock offsets with respect to the onboard GNSS receiver are derived from onboard GNSS observations. These serve as the reference for downlink signal modeling in subsequent steps.

Step 2: A batch least-squares estimation is applied to the downlink observations of LEO satellites to jointly estimate the downlink PCO and equipment delay parameters. Considering the short visible arcs of LEO downlink observations and the limited number of ground stations expected to be available within China in the foreseeable future, GNSS downlink observations are incorporated to assist in the estimation of ground station-related parameters such as receiver clock offset and zenith tropospheric delay (ZTD). This auxiliary information helps to decouple ground station parameters from satellite-specific parameters. In this step, inter-system bias (ISB) between the LEO and GNSS systems must also be estimated to maintain consistency in the joint observation model.

Step 3: Considering the temporal stability of the downlink PCO parameters for LEO satellites, the PCOs estimated in the second step are fixed. The corresponding equipment delays are used as initial values. A bidirectional filtering strategy under static PPP mode is then applied to further refine the equipment delay estimates. Subsequently, the estimation process can be simplified by executing only the first and third steps, allowing for continuous updates of the equipment delay to support long-term operational needs.

3. Simulation Results

Previous studies have demonstrated the feasibility of jointly estimating the downlink signal PCO and equipment delay using observations from single or a small number of LEO satellites. However, there remains a lack of systematic, comprehensive, and statistical evaluation regarding how estimation accuracy is affected by factors such as constellation size, orbital altitude, spatial-temporal error characteristics, and the number of ground stations. Therefore, this section aims to conduct a series of simulation-based experiments to assess the impact of these key factors on estimation performance, providing insights for future system design and deployment. Due to the unavailability of real-world observation data from ground receivers tracking the downlink signals of LEO satellites, simulation-based methods are adopted to generate representative datasets for the following experiments.

3.1. Data Simulation and Parameter Estimation Strategy

3.1.1. Constellation Simulation

To meet the requirements of the simulation experiments in this study, five polar LEO constellation configurations—labeled as Schemes A, B, C, D, and E were designed. The constellation parameters were inspired by the system architecture proposed in the paper for LEO-based navigation enhancement [14]. Detailed configuration parameters are listed in Table 1. These schemes span a range of satellite numbers, orbital altitudes, and distribution patterns, enabling a comprehensive evaluation of how constellation design affects the estimation accuracy of PCO and equipment delay.

Table 1. Detailed configuration parameters for LEO constellation.

Constellation Scheme	Constellation Type	Inclination	Number of Satellites	Orbital Altitude (km)	Satellites per Orbit Plane
A	Polar	-	144	500	12
B	Polar	-	144	1000	12
C	Polar	-	144	1500	12
D	Polar	-	72	1500	6
E	Polar	-	216	1500	18

Figure 4 illustrates the configuration of the simulated Scheme A LEO constellation. In this simulation, a polar LEO constellation is configured with 144 satellites evenly distributed across 6 orbital planes, with an approximate altitude of 1000 km. The LEO satellites are numbered sequentially as L01, L02, . . ., L144. For GNSS reference, the BDS-3 (BeiDou-3) constellation is adopted in the joint simulation and evaluation.

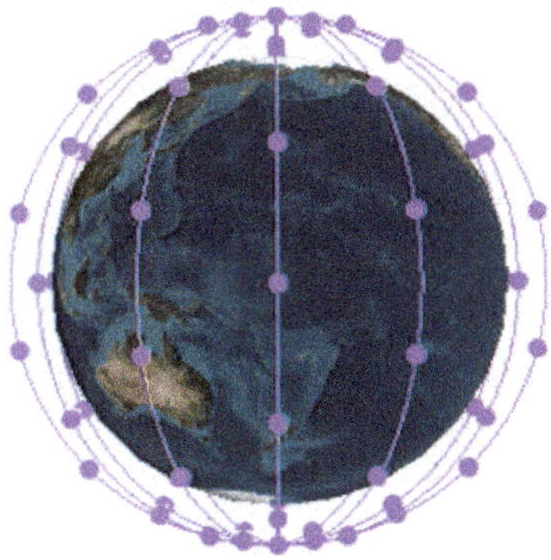

Figure 4. Configuration of the simulated LEO constellation.

To better approximate real-world conditions, this study adopts a realistic error modeling approach for the simulated LEO satellite orbits and clock biases. Orbital errors in LEO satellites are typically characterized by distinct periodic patterns, primarily resulting from imperfections in the dynamic force models, such as atmospheric drag and solar radiation pressure. In contrast, clock errors arise from a combination of systematic periodic variations (e.g., thermal drift or orbit-dependent effects) and stochastic noise introduced by the onboard oscillator. Accordingly, the orbital errors are modeled using deterministic periodic functions (e.g., cosine terms), while the clock errors are represented as a superposition of periodic components and Gaussian white noise [13]. This composite modeling strategy ensures a more faithful representation of orbit and clock bias behaviors, thereby enhancing the reliability of simulation-based performance evaluations. Figure 5 illustrates a representative example of the simulated orbit and clock bias errors.

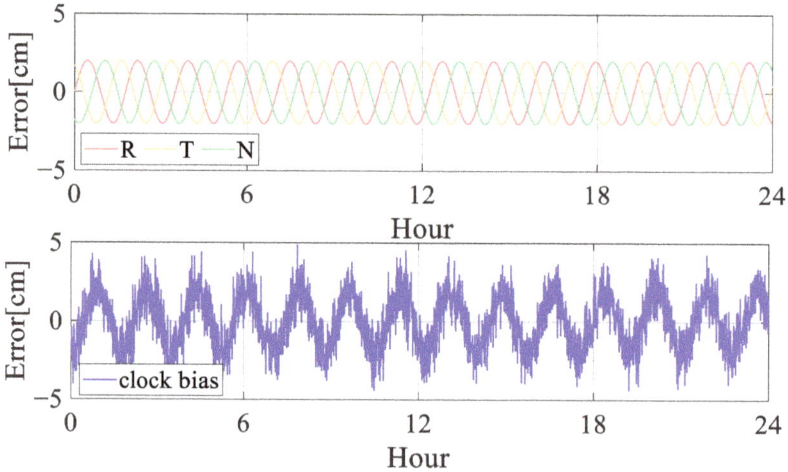

Figure 5. Simulation for the error of one LEO satellite orbit and clock bias error.

3.1.2. Observation Simulation and Ground Station Distribution

The ambiguities in phase observations linked to distinct receivers and frequencies are simulated as distinct integer ambiguities assigned to each satellite-frequency combination. The equipment delay parameters for LEO satellites are assumed as a constant ranging from tens to hundreds of nanoseconds, which vary with each satellite. In addition, the PCO corrections of GNSS satellites are simulated according to the precise ephemeris.

Based on the aforementioned configuration, the ionosphere-free downlink signal PCO parameters are defined in the satellite-fixed coordinate system, which follows the convention adopted by the GRACE-C satellite with the X-axis pointing along the flight direction, the Z-axis pointing toward the Earth's center, and the Y-axis completing the right-handed system. Different parameter values are assigned to individual LEO satellites to account for variations in onboard equipment delays.

The tropospheric delay is simulated according to the IGS precise ZTD parameters. The observation noises are set as zero-mean normal distributions with standard deviations of 0.3 m and 3 mm for pseudorange and phase observations, respectively. The other simulation configurations are outlined in Table 2.

Table 2. Simulation configurations of the observation data.

Error Source	Simulation Setting
Antenna phase center of ground receivers	Provided by IGS (ANTEX, GFZ German Research Centre for Geosciences, Potsdam, Germany)
Ground receiver clock offsets	Receiver clock offsets from IGS Multi-system PPP
IF equipment delay parameters for LEO satellite	Distinct equipment delay values for each satellite
ISB parameters for LEO	Each station is assigned distinct ISB values

Table 2. *Cont.*

Error Source	Simulation Setting
GPS/LEO satellites Ambiguity	A distinct constant integer for each phase observation of different satellite and frequency
Tropospheric error	ZTD parameter from IGS Multi-system PPP
Ionospheric error	Interpolated value according to GIM
Pseudorange noise	White noise with standard deviations of 0.3 m
Phase noise	White noise with standard deviations of 3 mm

In the initial stages of the development of the LEO satellite navigation system, it is conservatively estimated that the number of ground stations capable of tracking its navigation signals will be severely limited. This requires taking into account the scenario where global station deployment is not feasible and the number of available monitoring stations is scarce. Therefore, six network stations located in China are selected to simulate GPS and LEO observation shown in Figure 6, which are SHAO, URUM, CHAN, LHAZ, BJFS, and HKSL, respectively.

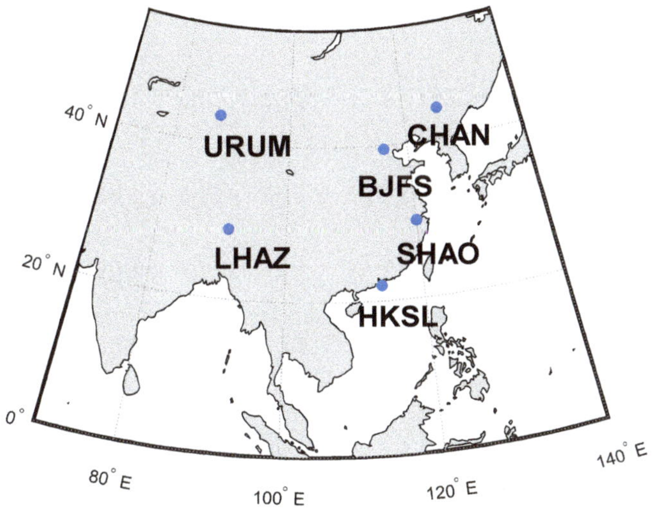

Figure 6. Distribution of Simulated Ground Stations in China.

3.1.3. Experimental Data and Processing Strategy

The configuration of precise ephemeris and observation data used for estimating the LEO downlink antenna phase center offset (PCO) and equipment delay is summarized in Tables 3 and 4. It is important to note that, in order to analyze the impact of different factors on the estimation of LEO downlink PCO and equipment delay, several parameters are provided with multiple configuration options. In addition, considering that a single LEO satellite typically passes over a ground station for only a few to several minutes—much shorter than the multi-hour visibility of GNSS satellites—a high-frequency observation interval of 5 s is adopted to enhance the utilization of observation data and improve the estimation accuracy.

Table 3. Configuration of Precise Orbit Ephemeris Parameters. (Unless otherwise specified, the baseline scheme is the one highlighted in bold and marked with an asterisk (*) in the table.).

Category	Configuration
Ephemeris Time Span	2 April 2023–8 April 2023
Sampling Interval	30 s
GNSS Constellation	BDS-3
LEO Constellation	A, B, **C** *, D, E
LEO Orbit Radial (R) Error Settings	**P2 cm** *, P3 cm
LEO Orbit Tangential (T) Error Settings	**P2 cm** *, P3 cm
LEO Orbit Normal (N) Error Settings	**P2 cm** *, P3 cm
LEO Clock Bias Error Settings	**P2 cm + W1 cm** * P3 cm + W1 cm P2 cm + W1.7 cm

For clarity and brevity, the following abbreviations are used in the tables to represent different types of simulated errors:

P: Periodic error (e.g., "P2 cm" denotes a periodic error with an amplitude of 2 cm)

W: White noise error (e.g., "W1 cm" denotes white noise with a standard deviation of 1 cm)

P + W: Combined periodic and white noise error (e.g., "P2 cm + W1 cm")

Table 4. Observation Data Configuration.

Category	Configuration
Sampling Interval	2 April 2023–8 April 2023
Sampling Interval	5 s
Observation Types	Simulated pseudorange and carrier phase data
Tracking System	BDS + LEO
Number of Stations in China	1, 2, 3, 4, 5, **6** *

Unless otherwise specified, the baseline scheme is the one highlighted in bold and marked with an asterisk (*) in the table.

The downlink signal PCO and equipment delay are estimated using a least-squares estimation (LSE) method based on combined BDS and LEO downlink observations. the estimated state vector includes the LEO downlink signal IF PCO, the IF equipment delay of the LEO satellite, the IF carrier phase ambiguities for both BDS and LEO satellites, the receiver clock bias, the receiver tropospheric delay, and the station ISB for LEO system (BDS as the reference). The overall estimation strategy is summarized in Table 5. After obtaining a reliable solution for the LEO antenna PCO, the PCO values are fixed, and estimation of the equipment delay is performed.

Table 5. Parameter Estimation Strategy.

Data Type	Instructions
Number of stations in China	**6** *
Data length	**7** * **day**
Observations	Pseudorange phase observation data of GPS satellites and LEO satellites
Signal	L1/L2
Sampling interval	5 s
Elevation Mask Angle	7°
Observation weight settings	Elevation angle weighting
Earth rotation parameters	ERP file provided by IERS
Tidal correction	Solid tide and extreme tide correction

Table 5. *Cont.*

Data Type	Instructions
Station coordinates	Weekly solution file in IGS products
Satellite orbit	BDS: IGS precision ephemeris; LEO: The orbit obtained by the combined orbit solution
Satellite clock	GPS: IGS precision clock; LEO: The clock obtained by the Combine solution
Tropospheric delay parameters	SAAS model correction+ WET delay walk estimation
Ionospheric parameters	IF combination
PCO\PCV of GPS satellites and Receiver	The igs14.atx file provided by IGS
Estimator	LSE
Parameters to be estimated	The receiver clock error The station tropospheric delay The BDS satellite IF ambiguity The LEO satellite IF ambiguity The LEO satellite downlink signals PCO under the satellite-fixed coordinate system The LEO satellite equipment delay The station ISB for LEO system (BDS as the reference)

3.2. Comparison of Parameter Correlations in Single LEO Satellite and LEO Constellation Modes

Since the correlation between parameters is a key factor affecting estimation accuracy, the parameter correlations in the context of LEO satellites are analyzed in detail. Two estimation modes are considered:

(1) Single LEO Satellite Mode: Only the downlink signal PCO and the corresponding equipment delay parameters of a single LEO satellite are estimated [30].

(2) LEO Constellation Mode: The downlink signal PCOs and equipment delays of all LEO satellites within the constellation are jointly estimated in a unified solution.

The correlation characteristics of the estimated parameters under both modes are compared to evaluate the impact of joint estimation across a constellation on parameter separability.

Figure 7 illustrates the mean correlation coefficient matrix of downlink signal PCO parameter, equipment delay parameter, LEO satellite ambiguity parameter, station clock parameter and station tropospheric delay parameters. The mean coefficient that is less than 1% indicates a low correlation between X direction of the downlink signal PCO and with other direction of PCO parameters. The mean correlation coefficients between Y direction and Z direction of the downlink signal PCO is approximately 47%. The mean correlation coefficients are less than 3% between LEO satellite equipment delay and other parameters. The mean correlation coefficients between LEO satellite ambiguity parameter with Y direction, Z direction of the downlink signal PCO parameter are approximately 24% and 33% respectively.

The correlation of estimated parameters in LEO satellite can be summarized as follows:

(1) The X direction of the downlink signal PCO and equipment delay parameters are basically independent of other parameters;

(2) The Y direction and Z direction of the downlink signal PCO have a strong correlation;

(3) LEO satellite ambiguity parameter has a closely correlation with the Y direction and Z direction of the downlink signal PCO parameter.

(4) The station estimated parameters are independent of the parameters to be estimated of LEO satellite.

Figure 8 illustrates the mean correlation coefficient matrix for LEO Constellation Mode. The result shows that while the overall correlation structure remains similar in both modes, the mean correlation coefficients are significantly reduced in the constellation mode. This suggests improved parameter separability, making the constellation-based estimation more favorable. Therefore, the LEO constellation mode is adopted in the subsequent analysis throughout this paper.

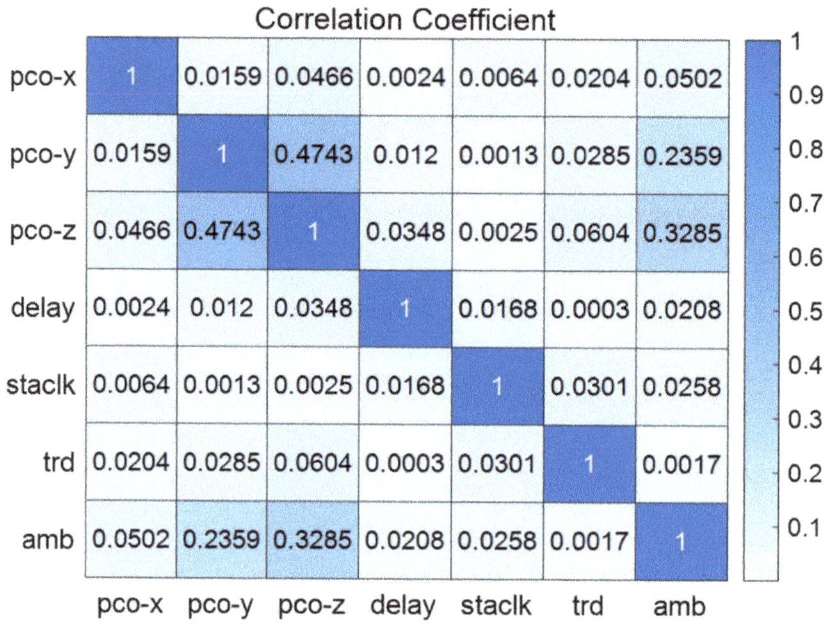

Figure 7. Correlation between the parameters of one LEO satellite. (pco-x, pco-y, and pco-z represent three directions of the downlink signal PCO of LEO satellites respectively; delay indicates the equipment delay of LEO satellites; staclk and trd represent station clock error and tropospheric delay parameters; amb represents the LEO satellite ambiguity parameter).

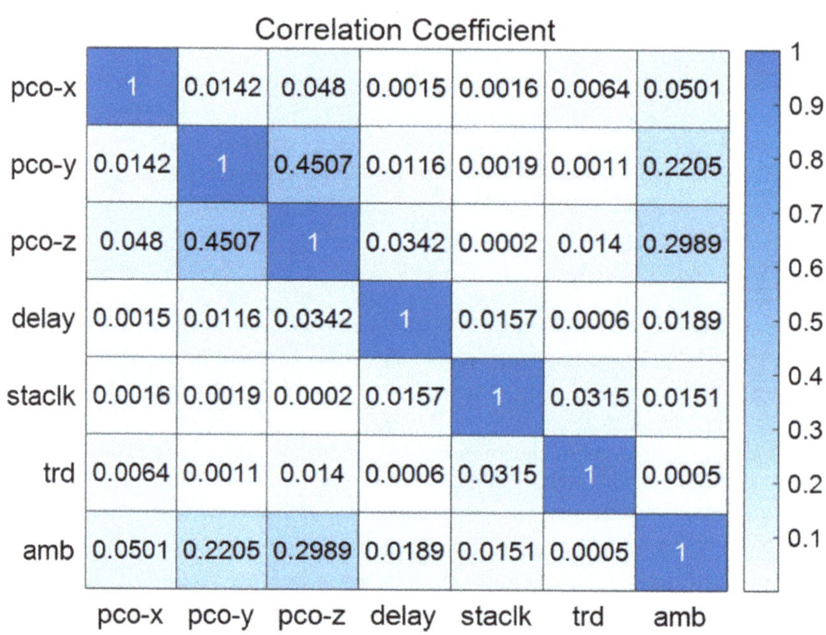

Figure 8. Correlation between the parameters of L12 satellite.

3.3. Comprehensive Evaluation of Factors Influencing LEO Downlink PCO and Equipment Delay Estimation

The feasibility of estimating LEO downlink signal PCO and equipment delay has been acknowledged. However, systematic, comprehensive, and statistically supported conclusions are still lacking with respect to the influence of key factors such as constellation size, orbital altitude, spatiotemporal accuracy requirements, and the number of ground stations on the estimation performance of these parameters. This section aims to perform a thorough evaluation of these factors to assess their respective impacts on the estimation accuracy, thereby providing valuable insights for the design and deployment of future LEO-based navigation augmentation systems.

To obtain a generalized assessment of the constellation-level estimation performance, statistical analysis is carried out based on the average PCO and equipment delay results across all satellites in the constellation. Simulation-based and statistical evaluations are used to support these findings. Specifically, the accuracy of PCO is assessed as the difference between the estimated value and its known true value from simulation, while the accuracy of the equipment delay is quantified by computing the root mean square (RMS) of the difference between the estimated and true values across all epochs.

3.3.1. Impact of Orbital Altitude on the Estimation Accuracy

Under a fixed constellation size, the impact of orbital altitude on the estimation accuracy of LEO downlink signal PCO and equipment delay is primarily reflected in two aspects: higher orbital altitudes result in more visible LEO satellites for ground stations, whereas lower altitudes lead to faster relative motion between satellites and the ground, enhancing the geometric variation but reducing the number of visible satellites. To investigate these effects systematically, three representative polar-orbit constellation schemes A, B, and C were selected, with orbital altitudes of 500 km, 1000 km, and 1500 km, respectively, and all containing 144 satellites. Figure 9 illustrates the estimation accuracy of PCO and equipment delay under LEO constellations with identical scale but different orbital altitudes. The corresponding statistical results are summarized in Table 6. The results show that the estimation accuracy in the X-direction is consistently higher than that in the Y and Z directions, with values better than 2 mm and relatively insensitive to changes in orbital height. This may be attributed to the alignment of the X-direction with the satellite velocity vector, which leads to faster changes in the corresponding partial derivatives in the design matrix and improves estimation sensitivity. In contrast, the Y and Z directions exhibit greater sensitivity to orbital altitude. At 500 km, the estimation accuracies of Y and Z directions are nearly equal; at 1000 km, the Y-direction performs worse than Z; while at 1500 km, the Z-direction accuracy deteriorates significantly. Regarding equipment delay, the best estimation accuracy is achieved at 1000 km, followed by 500 km, with the poorest performance at 1500 km. Nevertheless, all three configurations maintain equipment delay accuracy within 2 cm. Overall, these results confirm that orbital altitude has a significant impact on the estimation accuracy of LEO downlink PCO and equipment delay, through its influence on satellite visibility and geometric variation. Therefore, both factors should be carefully considered in constellation design and system implementation to optimize estimation performance.

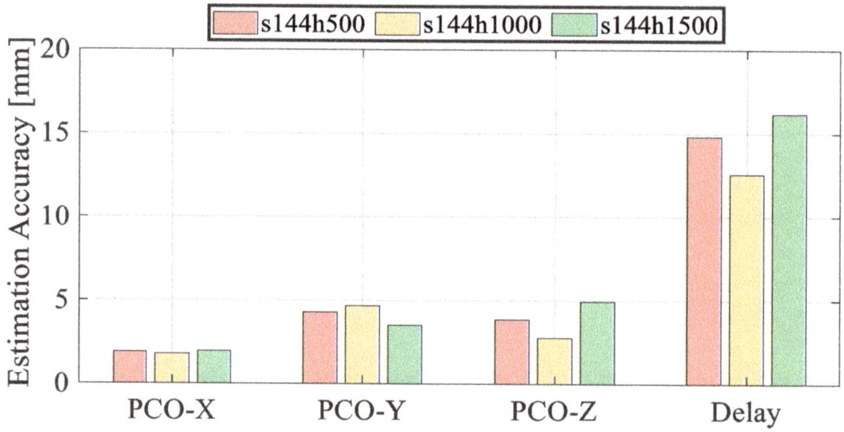

Figure 9. Comparison of PCO and Equipment Delay Estimation Accuracy for LEO Constellations with Identical Scale at Different Orbital Altitudes.

Table 6. Estimation Accuracy Statistics of PCO and Equipment Delay for Three Orbital Altitude Schemes (Unit: mm).

Constellation	PCO-X	PCO-Y	PCO-Z	Equipment Delay
LEO_s144h500	1.89	4.28	3.87	14.82
LEO_s144h1000	1.77	4.67	2.78	12.56
LEO_s144h1500	1.94	3.53	4.94	16.17

3.3.2. Impact of Constellation Scale on the Estimation Accuracy

To investigate the influence of constellation scale on the estimation accuracy of downlink signal PCO and equipment delay, three polar-orbit LEO constellation schemes C, D, and E were selected, consisting of 72, 144, and 216 satellites respectively. All constellations share the same configuration of 12 orbital planes and an altitude of 1500 km. The 72-satellite constellation can be extended to 144 or 216 satellites by progressively densifying the number of satellites per orbit. Figure 10 illustrates the estimation accuracy of PCO and equipment delay under different constellation scales, with detailed statistics provided in Table 7. As shown in Table 7, increasing the number of satellites in the constellation improves the estimation accuracy of both PCO and equipment delay. The X component of LEO downlink signal PCO remains relatively stable across configurations, consistently better than 3 mm. In contrast, the Y and Z directions show more substantial improvement—from 4.41 mm and 6.64 mm in the 72-satellite scheme to 3.40 mm and 4.68 mm in the 216-satellite configuration. Similarly, the equipment delay estimation improves from 20.01 mm to 15.65 mm as the satellite number increases. These results demonstrate that, under consistent orbital geometry, increasing the constellation scale enhances the observation redundancy and overall geometric strength, thereby improving the observability and estimation precision of key parameters.

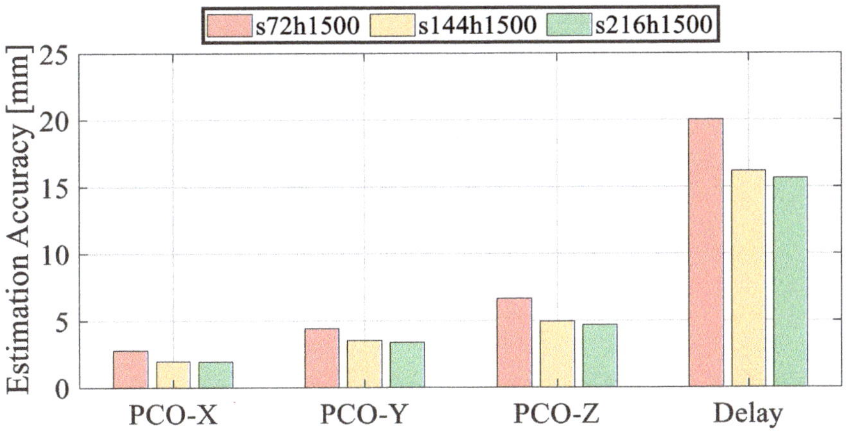

Figure 10. Comparison of PCO and Equipment Delay Estimation Accuracy for LEO Constellations of Different Scales.

Table 7. Estimation Accuracy of PCO and Equipment Delay under Different LEO Constellation Scales (Unit: mm).

Constellation	PCO-X	PCO-Y	PCO-Z	Equipment Delay
LEO_s72h1500	2.75	4.41	6.64	20.01
LEO_s144h1500	1.94	3.53	4.94	16.17
LEO_s216h1500	1.93	3.40	4.68	15.65

3.3.3. Impact of Orbit and Clock Bias Errors on the Estimation Accuracy

In practical precise orbit determination and satellite clock bias estimation, the results are inevitably affected by residual errors. Unmodeled orbit and clock errors in LEO satellites may be absorbed by estimable parameters such as the downlink antenna phase center offset (PCO) and equipment delay, thereby degrading the overall estimation accuracy. Therefore, the precision of orbit and clock products plays a critical role in the reliable estimation of downlink parameters. This section investigates the impact of orbit and clock errors on the estimation accuracy of PCO and equipment delay by simulating five combinations of error configurations. Figure 11 illustrates the estimation results under different error settings, with detailed statistics provided in Table 8.

Table 8. Estimation Accuracy of PCO and Equipment Delay under Different Orbit and Clock Bias Error Configurations (Unit: mm).

Scheme	Orbit R\T\N Error	Clock bias Error	PCO-X	PCO-Y	PCO-Z	Equipment Delay
1	P2 cm\P2 cm\P2 cm	P2 cm + W1 cm	1.94	3.53	4.94	16.17
2	P3 cm\P2 cm\P2 cm	P2 cm + W1 cm	2.16	5.02	8.06	20.85
3	P2 cm\P3 cm\P2 cm	P2 cm + W1 cm	3.05	4.03	6.07	18.64
4	P2 cm\P2 cm\P3 cm	P2 cm + W1 cm	2.40	4.64	6.36	19.37
5	P2 cm\P2 cm\P2 cm	P3 cm + W1 cm	2.72	5.13	5.69	21.73
6	P2 cm\P2 cm\P2 cm	P2 cm + W1.7 cm	2.21	4.31	5.20	23.42

A comparative analysis of all test schemes demonstrates that both orbit and clock errors significantly affect the estimation accuracy of LEO downlink parameters. Among the three components, the Z direction of the PCO is the most sensitive, with its estimation error increasing from 4.94 mm in Scheme 1 to 8.06 mm in Scheme 2 and 6.36 mm in

Scheme 4 when periodic errors are introduced in the radial (R) and normal (N) orbit directions, respectively. The equipment delay estimation is also notably degraded, reaching 23.73 mm under intensified white noise conditions (Scheme 6), representing an increase of approximately 47% compared to the best-case configuration. A comparison among Scheme 1,2 and Scheme 3, and 4 reveals that periodic errors in the R direction exert a more pronounced impact on estimation performance than those in the T or N directions. The contrast between Scheme 1, Scheme 5 and Scheme 6 further indicates that periodic errors in clock bias mainly degrade PCO estimation, while white noise errors predominantly affect the accuracy of equipment delay estimation. Additionally, the X component consistently exhibits the highest estimation accuracy across all schemes, indicating that its geometric sensitivity to orbit and clock errors is relatively low. In conclusion, achieving high-precision orbit and clock products is essential for improving the estimation accuracy of downlink antenna PCO and equipment delay.

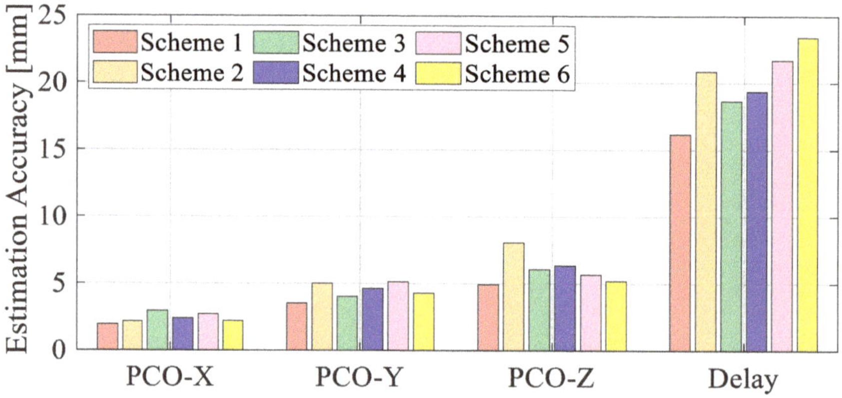

Figure 11. Comparison of PCO and Equipment Delay Estimation Accuracy under Various Orbit and Clock Error Configurations.

3.3.4. Impact of Ground Station Number and Processing Duration on Estimation Accuracy

To investigate the influence of ground station quantity and data processing duration on the estimation accuracy of downlink signal phase center offset (PCO) and equipment delay, a series of simulation experiments were designed with varying combinations of tracking stations and processing durations. Figure 12 illustrates the variation trends of estimation biases for the three PCO components and equipment delay under different numbers of stations (1–6) and processing durations (1–7 days). Tables 9 and 10 respectively summarize the results for fixed 7-day processing with varying station numbers, and for fixed three-station configurations with varying durations.

Table 9. Estimation Biases of PCO and Equipment Delay for Different Station Numbers with a Fixed 7-Day Processing Duration (Unit: mm).

Station Number	PCO-X	PCO-Y	PCO-Z	Equipment Delay
1	3.47	6.00	7.30	30.71
2	2.54	4.44	5.60	26.56
3	2.40	3.49	5.74	24.89
4	2.15	3.55	5.83	23.05
5	2.04	3.43	5.07	18.34
6	1.94	3.53	4.94	16.17

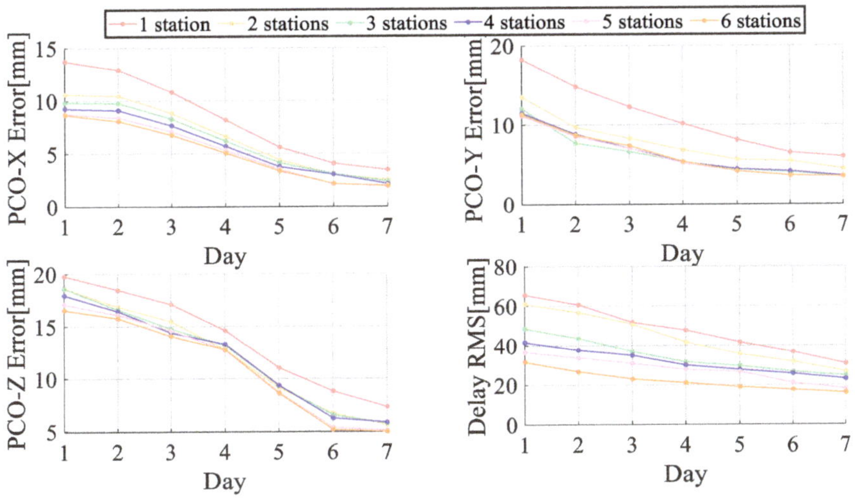

Figure 12. Estimation Accuracy of PCO and Equipment Delay Under Varying Station Numbers and Processing Durations.

Table 10. Estimation Biases of PCO and Equipment Delay for Different Processing Durations with Three Ground Stations (Unit: mm).

Processing Time (Day)	PCO-X	PCO-Y	PCO-Z	Equipment Delay
1	9.82	12.03	18.59	48.67
2	9.75	7.71	16.58	43.54
3	8.26	6.63	14.74	37.01
4	6.20	5.34	13.21	31.77
5	4.13	4.42	9.33	29.84
6	3.10	4.09	6.54	26.63
7	2.40	3.49	5.74	24.71

The results show that even with only one ground station, the 7-day processing configuration yields sub-centimeter estimation accuracy for PCO, demonstrating acceptable feasibility. Moreover, when the processing duration is held constant, increasing the number of ground stations leads to noticeable improvements in estimation performance. For instance, as shown in Table 9, increasing the station number from 1 to 3 reduces the X direction bias from 3.47 mm to 2.40 mm and the equipment delay bias from 30.71 mm to 24.89 mm. Further increases up to 6 stations continue to improve the estimates but with diminishing returns, indicating that 3–4 stations are sufficient to provide stable geometric coverage, while additional stations offer marginal improvements.

When the number of stations is fixed at three, extending the processing duration significantly enhances estimation accuracy, as shown in Table 10. From day 1 to day 7, estimation biases for all PCO components and equipment delay steadily decrease. During the first three days, the Y-direction PCO shows the most rapid improvement, while the X and Z directions exhibit more substantial gains from days 3 to 7, particularly the Z direction, whose bias decreases from 18.59 mm to 5.74 mm. The equipment delay estimate improves considerably within the first four days, with subsequent improvements tapering off, indicating that while longer observation periods improve observability, the marginal benefit diminishes beyond a certain point.

In summary, increasing the number of ground stations (preferably 3–4) and extending the processing time (ideally 5–7 days) can significantly improve the estimation accuracy of downlink PCO and equipment delay. The Z direction of PCO and equipment delay are particularly sensitive to these factors.

3.4. Discussion

The simulation results demonstrate that the proposed joint estimation framework can achieve sub-centimeter accuracy for PCO and centimeter-level accuracy for equipment delays, even with only 3–4 regionally distributed ground stations. Compared with previous studies [18,30], which mainly relied on single or a few experimental satellites under static scenarios, this work extends the evaluation to constellation-scale conditions, systematically analyzing the impacts of orbital altitude, constellation size, orbit/clock bias errors, and ground station configurations. These results highlight the practical applicability of the method for large-scale LEO navigation constellations.

A key finding is that the X-direction PCO component is consistently the most stable across different scenarios, while the Y- and Z-directions are more sensitive to orbital and clock errors. This observation aligns with the geometric characteristics of LEO orbits but also reveals new insights into parameter separability at the constellation level. Another important observation is the diminishing returns as constellation size increases: while additional satellites improve estimation accuracy, the marginal benefits decrease beyond a certain constellation scale. Similarly, increasing the number of ground stations improves parameter observability, but 3–4 well-distributed stations are generally sufficient to ensure robust performance.

Despite these encouraging results, several limitations should be acknowledged. First, the simulations are based on idealized assumptions of data availability and error modeling. In real-world scenarios, factors such as thermal deformation, structural instability, and long-term hardware drifts could impact PCO and equipment delay behavior. Second, the simplified modeling of orbit and clock errors may underestimate their influence on parameter estimation.

Future research should focus on validating the proposed framework using real LEO downlink observations. Additionally, integrating multi-frequency GNSS observations, assessing the long-term stability of LEO equipment delays, and developing real-time implementation strategies will be crucial to advancing the operational deployment of LEO-augmented navigation systems.

4. Conclusions

This study proposes a practical estimation framework for jointly determining the PCO and equipment delay of LEO satellite downlink signals, particularly tailored to scenarios where only a small number of regionally distributed ground stations are available within China. Through a series of simulations, the estimation accuracy was evaluated under various conditions, including different constellation sizes, orbital altitudes, spatiotemporal precision levels, numbers of ground stations, and processing durations, thereby validating the effectiveness and robustness of the proposed method. The main conclusions are summarized as follows:

(1) The proposed algorithm enables accurate joint estimation of LEO downlink signal PCO and equipment delay, achieving sub-centimeter accuracy for PCO and centimeter-level accuracy for equipment delay;

(2) Among the PCO components, the X-direction shows the most stable accuracy, while the Y and Z directions are more sensitive to orbit and clock bias errors—especially under low altitudes or sparse station distributions;

(3) Increasing the number of satellites improves estimation performance; however, the improvement diminishes beyond a certain constellation scale, showing a clear

marginal effect; Periodic components in orbit and clock bias errors exert significant impact on estimation accuracy, particularly affecting the Z-direction PCO and equipment delay;

(4) The observations of 3–4 regionally distributed ground stations over a 5–7 day observation period is sufficient to achieve stable and accurate estimation results.

These findings validate that the proposed method remains effective even under the constraint of a sparse regional ground network in China, highlighting its practicality for practical deployment.

Author Contributions: Z.L.: Conceptualization, Methodology, Software, Validation, Formal analysis, Investigation, Writing—original draft. W.L.: Supervision, Project administration, Funding acquisition, Writing—review & editing, Corresponding author. J.H.: Resources, Data curation, Visualization, Writing—review & editing. All authors have read and agreed to the published version of the manuscript.

Funding: This contribution is supported by the Key Research and Development Program of Hubei Province (Grant No. 2022BAA054), and the National Nature Science Foundation of China (No. 42388102).

Data Availability Statement: The observation data and satellite orbit and clock data used in the simulation were derived from publicly accessible datasets and in-house simulation tools.

Conflicts of Interest: The authors declare no conflicts of interest.

References

1. Zumberge, J.F.; Heflin, M.B.; Jefferson, D.C.; Watkins, M.M.; Webb, F.H. Precise point positioning for the efficient and robust analysis of GPS data from large networks. *J. Geophys. Res. Solid Earth* **1997**, *102*, 5005–5017. [CrossRef]
2. Kouba, J.; Héroux, P. Precise Point Positioning Using IGS Orbit and Clock Products. *GPS Solut.* **2001**, *5*, 12–28. [CrossRef]
3. Ge, M.; Gendt, G.; Rothacher, M.; Shi, C.; Liu, J. Resolution of GPS carrier-phase ambiguities in Precise Point Positioning (PPP) with daily observations. *J. Geod.* **2008**, *82*, 389–399. [CrossRef]
4. Xiao, G.; Li, P.; Gao, Y.; Heck, B. A Unified Model for Multi-Frequency PPP Ambiguity Resolution and Test Results with Galileo and BeiDou Triple-Frequency Observations. *Remote Sens.* **2019**, *11*, 116. [CrossRef]
5. Zhao, L.; Yang, Y.; Xiang, Z.; Zhang, S.; Li, X.; Wang, X.; Ma, X.; Hu, C.; Pan, J.; Zhou, Y.; et al. A Novel Low-Cost GNSS Solution for the Real-Time Deformation Monitoring of Cable Saddle Pushing: A Case Study of Guojiatuo Suspension Bridge. *Remote Sens.* **2022**, *14*, 5174. [CrossRef]
6. Li, B.; Ge, H.; Ge, M.; Nie, L.; Shen, Y.; Schuh, H. LEO enhanced Global Navigation Satellite System (LeGNSS) for real-time precise positioning services. *Adv. Space Res.* **2019**, *63*, 73–93. [CrossRef]
7. Ge, H.; Li, B.; Nie, L.; Ge, M.; Schuh, H. LEO constellation optimization for LEO enhanced global navigation satellite system (LeGNSS). *Adv. Space Res.* **2020**, *66*, 520–532. [CrossRef]
8. Deren, L.I.; Xin, S.; Jianya, G.; Jun, Z.; Jianhua, L.U. On Construction of China’s Space Information Network. *Geomat. Inf. Sci. Wuhan Univ.* **2015**, *40*, 711–715. [CrossRef]
9. Yang, Y. Concepts of Comprehensive PNT and Related Key Technologies. *Acta Geod. Cartogr.* **2016**, *45*, 505–510. [CrossRef]
10. Xie, J.; Kang, C. Engineering Innovation and the Development of the BDS-3 Navigation Constellation. *Engineering* **2021**, *7*, 558–563. [CrossRef]
11. Yang, Y.; Mao, Y.; Ren, X.; Jia, X.; Sun, B. Demand and key technology for a LEO constellation as augmentation of satellite navigation systems. *Satell. Navig.* **2024**, *5*, 11. [CrossRef]
12. Li, X.; Li, X.; Ma, F.; Yuan, Y.; Zhang, K.; Zhou, F.; Zhang, X. Improved PPP Ambiguity Resolution with the Assistance of Multiple LEO Constellations and Signals. *Remote Sens.* **2019**, *11*, 408. [CrossRef]
13. Li, X.; Ma, F.; Li, X.; Lv, H.; Bian, L.; Jiang, Z.; Zhang, X. LEO constellation-augmented multi-GNSS for rapid PPP convergence. *J. Geod.* **2019**, *93*, 749–764. [CrossRef]
14. Ma, F.; Zhang, X.; Li, X.; Cheng, J.; Guo, F.; Hu, J.; Pan, L. Hybrid constellation design using a genetic algorithm for a LEO-based navigation augmentation system. *GPS Solut.* **2020**, *24*, 62. [CrossRef]
15. Ke, M.; Lv, J.; Chang, J.; Dai, W.; Tong, K.; Zhu, M. Integrating GPS and LEO to accelerate convergence time of precise point positioning. In Proceedings of the 2015 International Conference on Wireless Communications & Signal Processing (WCSP), Nanjing, China, 15–17 October 2015; pp. 1–5. [CrossRef]

16. Ge, H.; Meng, G.; Li, B. LEO enhanced GNSS (LeGNSS) precise point positioning with emphasis on model comparison. *Adv. Space Res.* **2024**, *74*, 2156–2168. [CrossRef]
17. Li, X.; Lv, H.; Ma, F.; Li, X.; Liu, J.; Jiang, Z. GNSS RTK Positioning Augmented with Large LEO Constellation. *Remote Sens.* **2019**, *11*, 228. [CrossRef]
18. Li, W.; Yang, Q.; Du, X.; Li, M.; Zhao, Q.; Yang, L.; Qin, Y.; Chang, C.; Wang, Y.; Qin, G. LEO augmented precise point positioning using real observations from two CENTISPACE™ experimental satellites. *GPS Solut.* **2023**, *28*, 44. [CrossRef]
19. Chang, C.; Zhao, Q.; Li, M.; Li, W. Augmentation message design for LEO-enhanced precise positioning: In-orbit performance assessment. *Measurement* **2025**, *243*, 116314. [CrossRef]
20. Bock, H.; Jäggi, A.; Meyer, U.; Dach, R.; Beutler, G. Impact of GPS antenna phase center variations on precise orbits of the GOCE satellite. *Adv. Space Res.* **2011**, *47*, 1885–1893. [CrossRef]
21. Zhao, X.; Zhou, S.; Ci, Y.; Hu, X.; Cao, J.; Chang, Z.; Tang, C.; Guo, D.; Guo, K.; Liao, M. High-precision orbit determination for a LEO nanosatellite using BDS-3. *GPS Solut.* **2020**, *24*, 102. [CrossRef]
22. Luthcke, S.B.; Zelensky, N.P.; Rowlands, D.D.; Lemoine, F.G.; Williams, T.A. The 1-Centimeter Orbit: Jason-1 Precision Orbit Determination Using GPS, SLR, DORIS, and Altimeter Data Special Issue: Jason-1 Calibration/Validation. *Mar. Geod.* **2003**, *26*, 399–421. [CrossRef]
23. Nie, L.; Wang, J.; Tang, L.; Ge, H.; Ge, M.; Schuh, H. Impact of LEO configuration on GPS precise orbit determination with un-differenced ambiguity resolution. *GPS Solut.* **2025**, *29*, 28. [CrossRef]
24. Saroufim, J.; Kassas, Z.M. Ephemeris and Timing Error Disambiguation Enabling Precise LEO PNT. *IEEE Trans. Aerosp. Electron. Syst.* **2025**, *61*, 6138–6153. [CrossRef]
25. Allahvirdi-Zadeh, A.; Wang, K.; El-Mowafy, A. POD of small LEO satellites based on precise real-time MADOCA and SBAS-aided PPP corrections. *GPS Solut.* **2021**, *25*, 31. [CrossRef]
26. Mao, X.; Arnold, D.; Girardin, V.; Villiger, A.; Jäggi, A. Dynamic GPS-based LEO orbit determination with 1 cm precision using the Bernese GNSS Software. *Adv. Space Res.* **2021**, *67*, 788–805. [CrossRef]
27. Gong, X.; Zhang, W.; Wang, Q.; Wang, F.; Li, X.; Sang, J.; Liu, W. Precise real-time navigation of the small TJU-1 satellite using GPS, GLONASS and BDS. *Measurement* **2022**, *204*, 112090. [CrossRef]
28. Ge, H.; Meng, G.; Li, B. Zero-Reconvergence PPP for Real-Time Low-Earth Satellite Orbit Determination in Case of Data Interruption. *IEEE J. Sel. Top. Appl. Earth Obs. Remote Sens.* **2024**, *17*, 4705–4715. [CrossRef]
29. Montenbruck, O.; Hackel, S.; Wermuth, M.; Zangerl, F. Sentinel-6A precise orbit determination using a combined GPS/Galileo receiver. *J. Geod.* **2021**, *95*, 109. [CrossRef]
30. Liu, J.; Wang, K.; El-Mowafy, A.; Yang, X. PCO and hardware delay calibration for LEO satellite antenna downlinking navigation signals. *Meas. Sci. Technol.* **2024**, *35*, 086314. [CrossRef]
31. Hofmann-Wellenhof, B.; Lichtenegger, H.; Wasle, E. *GNSS–Global Navigation Satellite Systems: GPS, GLONASS, Galileo, and More*; Springer Science & Business Media: Berlin/Heidelberg, Germany, 2007.
32. Guo, F.; Zhang, X.; Wang, J. Timing group delay and differential code bias corrections for BeiDou positioning. *J. Geod.* **2015**, *89*, 427–445. [CrossRef]
33. An, X.; Meng, X.; Jiang, W. Multi-constellation GNSS precise point positioning with multi-frequency raw observations and dual-frequency observations of ionospheric-free linear combination. *Satell. Navig.* **2020**, *1*, 7. [CrossRef]
34. Wang, Z.; Li, Z.; Wang, L.; Wang, N.; Yang, Y.; Li, R.; Zhang, Y.; Liu, A.; Yuan, H.; Hoque, M. Comparison of the real-time precise orbit determination for LEO between kinematic and reduced-dynamic modes. *Measurement* **2022**, *187*, 110224. [CrossRef]
35. China Satellite Navigation Office (CSNO). BeiDou Navigation Satellite System Signal in Space Interface Control Document—BDS-3 Open Service Signal B1C/B2a/B2b (Version 1.0). Beijing, China. 2019. Available online: http://www.beidou.gov.cn (accessed on 1 December 2019).

remote sensing

MDPI

Article

A Continuous B2b-PPP Model Considering Interruptions in BDS-3 B2b Orbits and Clock Corrections as Well as Signal-in-Space Range Error Residuals

Rui Shang [1,†], Zhenhao Xu [1,†], Chengfa Gao [1,*], Xiaolin Meng [2], Wang Gao [2] and Qi Liu [1]

1 School of Transportation, Southeast University, Nanjing 211189, China; shangr@seu.edu.cn (R.S.); 220223035@seu.edu.cn (Z.X.); lq_0328@seu.edu.cn (Q.L.)
2 School of Instrument Science and Engineering, Southeast University, Nanjing 210096, China; xiaolin_meng@seu.edu.cn (X.M.); gaow@seu.edu.cn (W.G.)
* Correspondence: gaochfa@seu.edu.cn
† These authors contributed equally to this work.

Abstract: In 2020, BDS-3 began broadcasting high-precision positioning correction products through B2b signals, effectively addressing the limitations of ground-based augmentation. However, challenges such as the "south wall effect" from geostationary orbit (GEO) satellites, issues of data (IOD) mismatch, and signal priority conflicts often result in interruptions and anomalies during real-time positioning with the B2b service. This paper proposes a continuous B2b-PPP (B2b signal-based Precise Point Positioning) model that incorporates signal-in-space range error (SISRE) residuals and predictions for B2b orbits and clock corrections to achieve seamless, high-precision continuous positioning. In our experiments, we first analyze the characteristics of B2b SISRE for both BDS-3 and GPS. We then evaluate the positioning accuracy of several models, B2b-PPP, EB2b-PPP, PB2b-PPP, EB2bS-PPP, and PB2bS-PPP, through simulated and real dynamic experiments. Here, 'E' indicates the direct utilization of the previous observation corrections from B2b before the signal interruption, 'P' represents B2b prediction products, and 'S' signifies the incorporation of the SISRE residuals. The results show that EB2b-PPP exhibits significant deviations as early as 10 min into a B2b signal interruption. Both PB2b-PPP and EB2bS-PPP demonstrate comparable performances, with PB2bS-PPP emerging as the most effective method. Notably, in real dynamic experiments, PB2bS-PPP maintains positioning accuracy in the E/N directions like B2b-PPP, even after 40 min of signal interruption, ensuring continuous and stable positioning upon signal restoration. This achievement significantly enhances the capability for high-precision continuous positioning based on B2b signals.

Keywords: PPP-B2b service; PPP; orbits prediction; clock prediction; SISRE

1. Introduction

PPP (Precise Point Positioning) technology enables high-precision, absolute positioning globally by accurately correcting various error sources. Its key advantage lies in its flexibility and convenience, as it only requires a single receiver [1,2]. However, traditional PPP often depends on post-processing with precise products or real-time precise orbit and clock corrections broadcast by ground networks [3,4]. This reliance limits its widespread application in areas such as smart transportation, precision agriculture, and high-precision positioning in remote regions like deserts and at sea. Satellite-based augmentation services (SBASs), which broadcast precise products via satellites, have a reduced dependency

https://doi.org/10.3390/rs17040618

on ground infrastructure, thereby providing vital support for large-scale, high-precision PPP [5,6].

With the development of BDS-3 (Beidou Navigation Satellite System), China officially released the PPP-B2b interface document in 2020 [7]. This document specifies the use of B2b signals from three geostationary orbit (GEO) satellites (C59 to C61) within the BDS-3 constellation as the data broadcast channel. These signals transmit correction parameters for BDS-3 and Global Positioning Systems (GPSs), including precise satellite orbits and clock corrections relative to the broadcast ephemeris (BRDC), precise differential code bias (DCB), and the user range accuracy index, to users in China and the surrounding regions (from 75°E to 135°E longitude and 10°N to 55°N latitude), as well as in areas extending up to 1000 km above the Earth's surface [8]. By receiving these precise satellite orbits and clock corrections, users can modify the corresponding BRDC to generate a real-time precise ephemeris, enabling free real-time high-precision positioning, velocity measurements, and timing services [9–12]. Analyses and evaluations indicate that B2b-PPP (B2b signal-based Precise Point Positioning) can provide centimeter-level accuracy in static conditions and decimeter-level accuracy in dynamic conditions [13–15]. On 23 January 2023, Galileo High Accuracy Service (HAS) also announced the provision of high-precision orbit, clock, and other products for both GPS and Galileo through the Galileo E6B signal and network, enabling high-precision PPP on a global scale [16,17].

In SBAS-based dynamic high-precision positioning applications, the continuity of PPP depends on the orbit and clock products remaining within their "nominal validity period". Unlike the global service provided by Galileo HAS, the B2b service requires at least one visible GEO satellite to ensure observation continuity. However, since the GEO satellites are fixed in position over the equator (at 80°E, 110.5°E, and 140°E), users in the northern hemisphere, especially in high-latitude regions, experience lower satellite elevation angles [18,19]. This makes the signal more susceptible to obstruction by tall buildings, mountains, or vegetation, resulting in signal instability or loss, and leading to the "south wall effect" (the southern occlusion issue of GEO satellite services in the northern hemisphere) [20]. In terms of service integrity, the B2b broadcast's bandwidth is 500 bps. Due to the limited downlink bandwidth of the satellite signal, the B2b service prioritizes the broadcast of BDS-3 corrections, which may limit the broadcast of GPS corrections and affect data completeness [21]. Regarding signal matching accuracy, the receiver must strictly match the IOD (issue of data) of the B2b signal to ensure that the correct orbits and clock corrections are being used. Any IOD mismatch could result in the receiver using outdated or inconsistent data, thereby impacting positioning accuracy [22].

Therefore, in practical applications, it is crucial to minimize the risk of B2b-PPP system failure caused by B2b signal packet loss due to the "south wall effect", broadcast prioritization, and matching errors. The key lies in effectively detecting anomalies in the B2b signals and predicting B2b products during abnormal conditions [23]. In our previous research, we established a quadratic anomaly detection mechanism based on epoch difference and median absolute deviation, leveraging the timing characteristics of B2b orbits and clock corrections [23]. Consequently, this paper focuses on the effective prediction of B2b corrections. According to CSNO 2020, the nominal validity period of DCB is 86,400 s, the nominal validity time for satellite orbits corrections is 96 s, and for clock corrections, it is only 12 s [7]. Therefore, the prediction of B2b products primarily involves forecasting orbits and clock corrections [24]. Studies have shown that using the last epoch product before a B2b interruption can maintain positioning accuracy for up to 10 min, but this empirical threshold is insufficient in terms of usability and reliability for real-time positioning [25]. Our earlier research quantitatively analyzed the temporal characteristics of these corrections for BDS-3 and GPS in B2b, and we developed differentiated prediction models to address the discontinuity in B2b-PPP caused by anomalies. The experimental results indicate that the prediction accuracy is significantly affected by the continuous observation arc of B2b. Under typical conditions, the effective prediction duration can reach 20–30 min, during which time the positioning accuracy is better than 0.1 m [23]. However, as time progresses, the predicted values gradually deviate from the true values, leading to a decline in positioning accuracy. Additionally, after long signal interruptions, when the signal is restored, significant differences between the predicted and real-time orbits/clock corrections can result in substantial positioning jumps.

The decline in precision and the positioning jumps observed during the reacquisition of B2b signals are primarily due to the degradation of the accuracy of the forecast products over time, leading to significant discrepancies from the actual B2b products. To address these issues, it is crucial to absorb these residual errors effectively. SISRE (signal-in-space range error), as a comprehensive metric, is typically used to reflect the accuracy of orbits and clock products or to indicate the overall error in satellite signals [26]. Therefore, SISRE parameters can be incorporated into positioning to absorb residual errors and enhance accuracy. Existing studies have demonstrated that incorporating SISRE parameters into BRDC-based high-precision PPP improves accuracy [27,28]. Similarly, studies indicate that for B2b signals, the SISRE standard deviation (STD) values range from 1.6 cm to 30.5 cm for BDS-3, with an average of 4.9 cm. For GPS, the values vary between 2.5 cm and 33.0 cm, with an average of 7.1 cm. Introducing SISRE parameters into B2b-PPP can effectively absorb residual errors in B2b orbits and clock corrections, thereby improving B2b-PPP accuracy [22].

In this paper, we propose incorporating SISRE parameters to effectively absorb forecast residuals based on predicted orbits and clock corrections. Additionally, recognizing the variations in predicted orbit and clock corrections accuracy among different satellites or satellite pairs, we have developed a stochastic model that incorporates SISRE accuracy. By reconstructing the B2b-PPP observation model and the stochastic model, we aim to achieve continuous and stable positioning during B2b signal interruptions. In the following sections, we will first provide a detailed introduction to the mathematical models. During the experiments, we analyzed the statistical characteristics of SISRE and validated the effectiveness of our proposed models through simulated kinematic positioning experiments and real-world kinematic tests. Finally, we present our conclusions.

2. Materials and Methods

In this section, we will first introduce the methodology for reconstructing precise ephemerides based on B2b signals. Next, we will present our previously proposed models for B2b orbits and clock predictions. Additionally, we will outline the assessment method for SISRE and formulate a B2b-PPP observation model along with a stochastic model that incorporates SISRE residuals.

2.1. Precise Orbits and Clock Generation Based on B2b Corrections

Currently, the B2b signal primarily broadcasts products including orbits and clock corrections for BDS-3 and GPS satellites, as well as DCB products for BDS-3. The orbit corrections refine the BRDC to obtain precise real-time satellite coordinates. These corrections include radial, along-track, and cross-track components and must be converted to the Earth Centered Earth Fixed (ECEF) coordinate system before application. For clock corrections, B2b directly refines the BRDC clock offset. The method for generating precise real-time B2b orbits and clock products is as follows [7]:

$$\begin{cases} \mathbf{X}^s_{prec} = \mathbf{X}^s_{brdc} + \begin{bmatrix} \mathbf{e}_{radial} & \mathbf{e}_{along} & \mathbf{e}_{cross} \end{bmatrix} \cdot \delta\mathbf{O} \\ \mathbf{e}_{radial} = \frac{\mathbf{r}}{|\mathbf{r}|}, \ \mathbf{e}_{cross} = \frac{\mathbf{r}\times\dot{\mathbf{r}}}{|\mathbf{r}\times\dot{\mathbf{r}}|}, \ \mathbf{e}_{along} = \mathbf{e}_{cross} \times \mathbf{e}_{radial} \\ \delta\mathbf{O} = \begin{bmatrix} d\mathbf{O}_{radial} & d\mathbf{O}_{along} & d\mathbf{O}_{cross} \end{bmatrix}^T \\ dt^s_{prec} = dt^s_{brdc} - dclk/c \end{cases} \tag{1}$$

where $\mathbf{X}^s_{prec}$ represents the real-time precise satellite position vector after applying B2b orbit corrections to the BRDC, and $\mathbf{X}^s_{brdc}$ is the satellite position vector calculated from the BRDC. The unit vectors $\mathbf{e}_{radial}$, $\mathbf{e}_{along}$, and $\mathbf{e}_{cross}$ correspond to the radial, along-track, and cross-track directions, respectively. The vector $\delta\mathbf{O}$ is the B2b orbit corrections vector in the ECEF coordinate frame; $d\mathbf{O}_{radial}$, $d\mathbf{O}_{along}$, and $d\mathbf{O}_{cross}$ represent B2b orbit corrections in the radial, along-track, and cross-track directions, respectively. Furthermore, $\mathbf{r}$ and $\dot{\mathbf{r}}$ are the satellite's position and velocity vectors in the ECEF coordinate system. The term dt^s_{prec} donates the corrected real-time precise satellite clock offset, while dt^s_{brdc} is the satellite clock offset calculated from the BRDC, $dclk$ represents the B2b clock corrections, and c is the speed of light in the vacuum.

2.2. B2b Orbits and Clock Corrections Prediction Model

Our preliminary research indicates that BDS-3 orbit corrections typically remain within 0.2 m and do not vary significantly. Therefore, when a signal interruption occurs, it is recommended to directly use the value from the last epoch as the predicted correction. In contrast, GPS orbits corrections exhibit larger variations and more frequent jumps. Instead, we employ a quadratic polynomial to predict GPS orbit corrections. For clock corrections, long-term observational data reveal that BDS-3 clock corrections follow a pattern more aligned with a linear polynomial. The variations in GPS satellites are more complex, sometimes showing strong stability while at other times exhibiting high-frequency fluctuations. Due to this irregularity, the average value of the last few epochs is used as the predicted correction. In summary, during signal interruptions, the following prediction models are applied for B2b orbits and clock corrections [23]:

$$\begin{cases} dX^C = dX^C_{last} \\ dX^G = a_0 + a_1\Delta t + a_2\Delta t^2 \\ dclk^C = b_0 + b_1\Delta t \\ dclk^G = \frac{1}{N}\sum_{i=0}^{N} dclk^G_{last-i} \end{cases} \tag{2}$$

where C and G represent BDS-3 and GPS, respectively, while dX^C and dX^G denote the predicted orbits corrections, and $dclk^C$ and $dclk^G$ denote the predicted clock corrections. The subscript $last$ indicates the last observation epoch, $last - i$ represents the previous i epochs, and Δt is the difference between the predicted epoch and the reference epoch. a_0 and a_1 are polynomial coefficients for GPS orbit prediction; b_0 and b_1 are polynomial coefficients for BDS-3 clock prediction. In practical applications, the polynomial coefficients can be fitted using the least squares method.

2.3. Assessment of B2b SISRE for BDS-3 and GPS

To evaluate the quality of B2b orbits and clock products, the international GNSS service (IGS) final precise ephemeris is commonly used as a reference for analyzing the accuracy of the corrected satellite orbits and clock data. SISRE quantifies the overall impact of orbits and clock errors on the user's line of sight, serving as a key metric for GNSS public services [26,28]. The SISRE for BDS and GPS can be calculated by combining the radial, along-track, and cross-track components of orbit errors with clock errors, as follows:

$$SISRE = \begin{cases} \sqrt{(0.98d_{radial} - c\delta t^s)^2 + \left(d_{along}^2 + d_{cross}^2\right)/54} & BDS(MEO) \\ \sqrt{(0.99d_{radial} - c\delta t^s)^2 + \left(d_{along}^2 + d_{cross}^2\right)/127} & BDS(IGSO) \\ \sqrt{(0.98d_{radial} - c\delta t^s)^2 + \left(d_{along}^2 + d_{cross}^2\right)/49} & GPS \end{cases} \quad (3)$$

where d_{radial}, d_{along}, and d_{cross} represent the radial, along-track, and cross-track orbits error, respectively, while δt^s denotes the clock error.

It is important to note that when calculating orbit errors, the B2b orbit corrections for BDS-3 are centered on the antenna phase center (APC) of the B3I frequency, while for GPS, they are centered on the APC of the L1/L2 ionosphere-free (IF) combination. In contrast, the satellite coordinates of the final precise products are based on the satellite's center of mass (CoM). Consequently, BDS-3 satellites require corrections using the phase center offset (PCO) products provided by CSNO, whereas GPS satellites need corrections based on the igs14.atx file corresponding to the relevant GPS week [29,30]. These adjustments eliminate discrepancies between the final precise orbits and B2b orbit products, facilitating the calculation of differences between the satellite orbit products.

When comparing B2b clock products with precise clock products, it is essential to consider the differences in frequency references and time scales. The reference signals used by IGS's precise clock products are based on a dual-frequency IF combination. Specifically, GPS utilizes the L1/L2 IF combination as its reference signal, which aligns with the B2b reference signals. In contrast, the precise clock products of BDS-3 from IGS typically use the B1I/B3I IF combination as their reference signal, with the B2b clock corrections referenced to B3I. To standardize the clock offset benchmarks across different products, appropriate DCB corrections are necessary. This allows for the calculation of the single-difference (SD) clock offset between B2b and the precise clock products. Additionally, systematic errors in the time scales of different products, along with fluctuations in various time references, can introduce certain systematic biases into the SD satellite clock offsets for each system. To address this issue, a double-difference (DD) approach is commonly employed, involving the subtraction of the SD clock offset of a reference satellite [31]. After evaluating the orbits and clock errors, the SISRE can be calculated using Equation (3).

2.4. B2b-PPP Observation Model Incorporating SISRE Residuals

The IF model is one of the most widely used observation models for PPP. It takes advantage of the inverse square relationship between ionospheric delay and frequency, using dual-frequency observations to eliminate the influence of the first-order ionospheric term. This section first outlines the basic observation model for PPP and then constructs the IF B2b-PPP model. Building on this foundation, an improved observation model is introduced by incorporating the SISRE residuals. The basic observation model is as follows [32]:

$$\begin{cases} P_{r,j}^s = \rho_r^s + cdt_r - cdt^s + M_r^s T_w + I_{r,j}^s + \left(d_{r,j} - d_j^s\right) + e_{r,j}^s \\ \varphi_{r,j}^s = \rho_r^s + cdt_r - cdt^s + M_r^s T_w - I_{r,j}^s + \left(\lambda_j N_{r,j}^s + b_{r,j} - b_j^s\right) + \varepsilon_{r,j}^s \end{cases} \quad (4)$$

where P and φ are the pseudo-range and carrier phase measurement, s and r represent the satellite and receiver, respectively, and ρ denotes the geometric distance between them. j is the corresponding carrier frequency. dt_r and dt^s represent the clock errors in the satellite and receiver, while M_r^s and T_w correspond to the mapping function for zenith tropospheric wet delay and the associated zenith tropospheric wet delay. $I_{r,j}^s$ represents the ionospheric delay, and $d_{r,j}$ and d_j^s denote the pseudo-range hardware delays for the receiver and satellite. $b_{r,j}$ and b_j^s are the carrier phase hardware delays for the receiver and satellite, with λ_j representing the corresponding carrier wavelength and $N_{r,j}^s$ representing the ambiguity. $e_{r,j}^s$ and $\varepsilon_{r,j}^s$ denote the pseudo-range and carrier phase noise, respectively. Other modellable errors, such as relativistic effects, Earth rotation effects, satellite antenna PCO and phase center variations (PCVs), phase wrapping, solid tide effects, and dry tropospheric delay, can be corrected using corresponding models, which are not detailed here.

In the B2b service, BDS-3 satellite clock corrections are referenced to the civil navigation (CNAV1) BRDC products, using the B3I frequency as the baseline. For GPS satellite clock corrections, the reference is the L1/L2 IF combination provided by the legacy navigation (LNAV) BRDC. Therefore, for users of the IF PPP service utilizing the B2b service, the DCB for the corresponding frequency points of the BDS-3 satellites should be corrected. In contrast, GPS satellites typically use the L1/L2 IF combination, so no DCB corrections are necessary for the B2b-PPP of GPS. Consequently, the BDS-3 IF model after parameter reorganization is structured as follows:

$$\begin{cases} P_{r,\mathrm{IF}_{ij}}^{s,C} = \rho_r^s + c\bar{d}t_r^C - c\bar{d}t^{s,C} + M_r^{s,C}T_w + \left(d_{B3I}^{s,C} - d_{\mathrm{IF}_{ij}}^{s,C}\right) + e_{r,\mathrm{IF}_{ij}}^{s,C} \\ \varphi_{r,\mathrm{IF}_{ij}}^{s,C} = \rho_r^s + c\bar{d}t_r^C - c\bar{d}t^{s,C} + M_r^{s,C}T_w + \lambda_{\mathrm{IF}_{ij}}\overline{N}_{r,\mathrm{IF}_{ij}}^{s,C} + \varepsilon_{r,\mathrm{IF}_{ij}}^{s,C} \end{cases} \tag{5}$$

where IF represents the IF combination, and i and j denote the indices of the two frequencies involved in the calculation; the receiver clock error, satellite clock error, and the floating ambiguity are reparametrized as follows:

$$\begin{cases} c\bar{d}t_r^C = cdt_r^C + d_{r,\mathrm{IF}_{ij}}^C \\ c\bar{d}t^{s,C} = cdt^{s,C} + d_{B3I}^s \\ \lambda_{\mathrm{IF}_{ij}}\overline{N}_{r,\mathrm{IF}_{ij}}^{s,C} = \lambda_{\mathrm{IF}_{ij}}N_{r,\mathrm{IF}_{ij}}^{s,C} + \left(b_{r,\mathrm{IF}_{ij}}^C - b_{\mathrm{IF}_{ij}}^{s,C}\right) - \left(d_{r,\mathrm{IF}_{ij}}^C - d_{B3I}^s\right) \end{cases} \tag{6}$$

In the pseudo-range observation model, the corresponding pseudo-range hardware delay bias can be corrected using the DCB products broadcast by the B2b service:

$$\begin{aligned} d_{\mathrm{IF}_{ij}}^{s,C} - d_{B3I}^{s,C} &= \left(\frac{f_i^2}{f_i^2-f_j^2}d_i^{s,C} - \frac{f_j^2}{f_i^2-f_j^2}d_j^{s,C}\right) - d_{B3I}^{s,C} \\ &= \frac{f_i^2}{f_i^2-f_j^2}\left(d_i^{s,C} - d_{B3I}^{s,C}\right) - \frac{f_j^2}{f_i^2-f_j^2}\left(d_j^{s,C} - d_{B3I}^{s,C}\right) \\ &= \alpha_{ij}DCB_{i-B3I}^{s,C} + \beta_{ij}DCB_{j-B3I}^{s,C} \end{aligned} \tag{7}$$

where $DCB_{i-B3I}^{s,C}$ and $DCB_{j-B3I}^{s,C}$ represent the DCB corrections for frequencies i and j, which can be obtained from the DCB corrections based on the corresponding identifiers. For GPS, the methods for reparametrizing receiver clock errors and carrier phase ambiguities are identical to those used for BDS-3 and will not be repeated here. Consequently, after applying the B2b corrections, the IF combination model for the BDS-3/GPS dual system is as follows:

$$\begin{cases} P_{r,\mathrm{IF}_{ij}}^{s,C} = \rho_r^s + c\bar{d}t_r^C + M_r^{s,C}T_w + e_{r,\mathrm{IF}_{ij}}^{s,C} \\ \varphi_{r,\mathrm{IF}_{ij}}^{s,C} = \rho_r^s + c\bar{d}t_r^C + M_r^{s,C}T_w + \lambda_{\mathrm{IF}_{ij}}\overline{N}_{r,\mathrm{IF}_{ij}}^{s,C} + \varepsilon_{r,\mathrm{IF}_{ij}}^{s,C} \\ P_{r,\mathrm{IF}_{ij}}^{s,G} = \rho_r^s + c\bar{d}t_r^G + M_r^{s,G}T_w + e_{r,\mathrm{IF}_{ij}}^{s,G} \\ \varphi_{r,\mathrm{IF}_{ij}}^{s,G} = \rho_r^s + c\bar{d}t_r^G + M_r^{s,G}T_w + \lambda_{IF_{ij}}\overline{N}_{r,\mathrm{IF}_{ij}}^{s,G} + \varepsilon_{r,\mathrm{IF}_{ij}}^{s,G} \end{cases} \tag{8}$$

The parameters to be estimated for the IF combination B2b-PPP of the BDS-3/GPS dual system can be written as:

$$\mathbf{X} = \left[dx, dy, dz, \bar{d}t_r^C, \bar{d}t_r^G, T_w, \overline{\mathbf{N}}_{r,\mathrm{IF}_{ij}}^{s,C}, \overline{\mathbf{N}}_{r,\mathrm{IF}_{ij}}^{s,G}\right] \tag{9}$$

where dx, dy, dz donates the three-dimensional coordinate vector, $\bar{d}t_r^C$ and $\bar{d}t_r^G$ represent the receivers clock bias after reorganization of BDS-3 and GPS, and $\overline{\mathbf{N}}_{r,\mathrm{IF}_{ij}}^{s,C}$ and $\overline{\mathbf{N}}_{r,\mathrm{IF}_{ij}}^{s,G}$ refer to the IF ambiguities after the reorganization of BDS-3 and GPS. During the prediction of orbits and clock errors, the errors tend to increase over time. This can be compensated for by introducing SISRE parameter $S_r^{s,C}$ and $S_r^{s,G}$ into the observation equation for each satellite, thereby correcting the orbits and clock errors. Considering the residuals of SISRE, the IF combination observations for the BDS-3/GPS are as follows:

$$\begin{cases} P_{r,\mathrm{IF}_{ij}}^{s,C} = \rho_r^s + c\bar{d}t_r^C + M_r^{s,C}T_w + S_r^{s,C} + e_{r,\mathrm{IF}_{ij}}^{s,C} \\ \varphi_{r,\mathrm{IF}_{ij}}^{s,C} = \rho_r^s + c\bar{d}t_r^C + M_r^{s,C}T_w + \lambda_{\mathrm{IF}_{ij}}\overline{N}_{r,\mathrm{IF}_{ij}}^{s,C} + S_r^{s,C} + \varepsilon_{r,\mathrm{IF}_{ij}}^{s,C} \\ P_{r,\mathrm{IF}_{ij}}^{s,G} = \rho_r^s + c\bar{d}t_r^G + M_r^{s,G}T_w + S_r^{s,G} + e_{r,\mathrm{IF}_{ij}}^{s,G} \\ \varphi_{r,\mathrm{IF}_{ij}}^{s,G} = \rho_r^s + c\bar{d}t_r^G + M_r^{s,G}T_w + \lambda_{\mathrm{IF}_{ij}}\overline{N}_{r,\mathrm{IF}_{ij}}^{s,G} + S_r^{s,G} + \varepsilon_{r,\mathrm{IF}_{ij}}^{s,G} \end{cases} \tag{10}$$

After linearization, the parameters can be estimated for the IF combination PPP of the BDS-3/GPS dual system, considering SISREs as follows:

$$\mathbf{X} = \left[dx, dy, dz, \bar{d}t_r^C, \bar{d}t_r^G, T_w, \mathbf{S}_r^{s,C}, \mathbf{S}_r^{s,G}, \overline{\mathbf{N}}_{r,\mathrm{IF}_{ij}}^{s,C}, \overline{\mathbf{N}}_{r,\mathrm{IF}_{ij}}^{s,G}\right] \tag{11}$$

Compared to the conventional IF PPP model, this model incorporates the SISRE parameter. During the solution process, the initial SISRE values are typically given based on empirical data, and they are estimated using a random walk approach [22,27]. However, considering the continuity of positioning, when both real-time B2b products and predicted B2b products are used in combination, more careful consideration is needed for the process noise in the random walk. This will be discussed in the next section.

2.5. Stochastic Model of SISRE

The random walk estimation method for SISRE can be expressed as follows [22,33]:

$$SISRE_{r,0}^s(k) = SISRE_r^s(k-1) + \omega_{SISRE_{r,0}^s(k)}, \omega_{SISRE_{r,0}^s(k)} \sim N\left(0, \sigma_{\omega^2_{SISRE_{r,0}^s(k)}}\right) \tag{12}$$

$$\sigma^2_{SISRE_{r,0}^s(k)} = \sigma^2_{SISRE_r^s(k-1)} + \sigma^2_{\omega^2_{SISRE_{r,0}^s(k)}} \tag{13}$$

where k represents the epoch number, $SISRE_{r,0}^s(k)$ is the initial value of SISRE at epoch k, $SISRE_r^s(k-1)$ is the estimated SISRE value at epoch $k-1$, and $\omega_{SISRE_{r,0}^s(k)}$ is the change in the SISRE parameter. $\sigma^2_{SISRE_{r,0}^s(k)}$ is the prior variance in SISRE at epoch k, $\sigma^2_{SISRE_r^s(k-1)}$ is the updated variance in the SISRE parameter at epoch $k-1$, and the variance corresponding to $\omega_{SISRE^s_{r,0}(k)}$ is $\sigma^2_{\omega^2_{SISRE^s_{r,0}(k)}}$.

When the B2b signal is being received normally, SISRE fluctuations tend to be relatively stable. The process noise from random walk is typically assumed to be constant, such as 1.2 cm/$\sqrt{h}$ for BDS-3 and 1.8 cm/$\sqrt{h}$ for GPS (h represents hours). This means that the process noise is related to the time interval from the previous epoch [22]. However, when using predicted B2b orbit and clock products, over time, the predicted values may begin to diverge from the actual values, leading to an increase in spatial signal residual errors. Moreover, following extended signal outages, when the signal is restored, large discrepancies between the predicted and actual orbits/clock corrections can cause substantial positioning jumps. Therefore, when using predicted products, it is necessary to reconsider the random walk estimation process for SISRE. It is important to note that due to the lack of real-time reference products, SISREs' accuracy cannot be assessed in real time. Therefore, we obtain the corresponding empirical values through the analysis of long-term sequences for each satellite.

3. Analysis of SISRE Under Different Interruption Scenarios

To analyze the accuracy of SISREs, we placed a ComNav AT360 antenna, manufactured by ComNav Technology Ltd. (located in Shanghai, China), on the rooftop of the School of Transportation at Southeast University, as illustrated in Figure 1. A ComNav K823 receiver board, also manufactured by ComNav Technology Ltd., was utilized to receive B2b data in real time. Data collection was conducted using STRSVR software (version 2.4.3) from RTKLIB [34], while decoding was performed with PPPB2bMsgDecode (version 1.0) software provided by ComNav Technology Ltd. We collected three days of B2b data from DOY 340 to 342 in 2023, using the final orbit and clock products from Wuhan University (WHU) as our reference. In this analysis, we assessed the accuracy of two types of post-interruption products: the first type being model-predicted B2b (Pred-B2b) products and the second type which uses correction values from the last observation epoch before the interruption as the predicted product (Extended B2b, Ext-B2b).

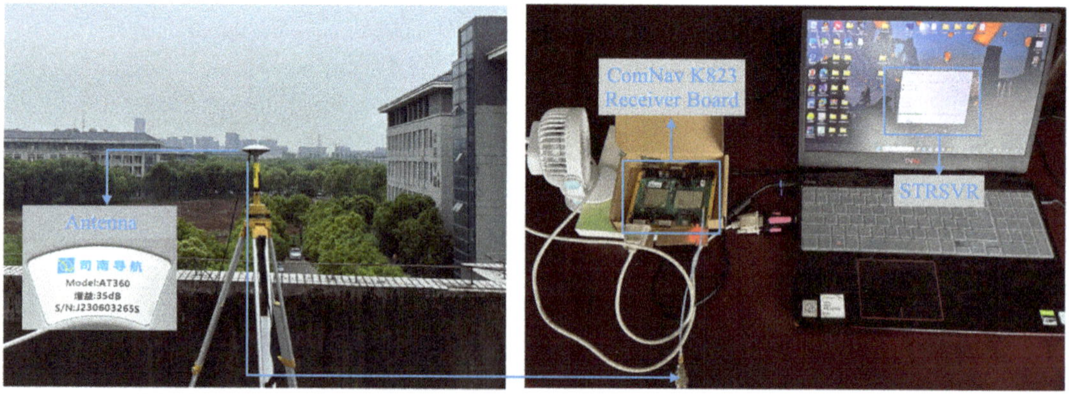

Figure 1. B2b data collection experimental setup. The **left panel** is the antenna we used, while the **right panel** displays the ComNav K823 receiver board.

Taking 8 h of data from GPS between the times 00:00 and 07:00 on DOY (Day of Year) 340 in 2023 as an example, we manually interrupted the B2b signal data at the 20th minute of each hour and then predicted the data for the subsequent time period (e.g., if the interruption occurred at 00:20, the data from 00:20 to 01:00 was Pred-B2b and Ext-B2b).

Figures 2 and 3 illustrate the SISRE of some of the BDS-3 and GPSs during each hour, comparing the continuous B2b arcs under normal conditions with those under Pred and Ext scenarios. As is evident from Figure 2, for BDS-3, the SISREs are generally less than 0.1 m. From 02:00 to 03:00, for C32, the Pred-B2b aligns well with B2b, whereas the Ext-B2b exhibits a noticeable deviation. For C20 and C39 from 06:00 to 08:00, although both Pred-B2b and Ext-B2b show deviations, Pred-B2b clearly demonstrates a better fit. Turning to Figure 3, the GPS exhibits similar characteristics; however, as shown in Hour 4, the SISRE of B2b, along with the SISRE of the Pred-B2b and Ext-B2b, exceeds 0.15 m. This is attributed to the relatively low accuracy of the B2b corrections in the GPS [30].

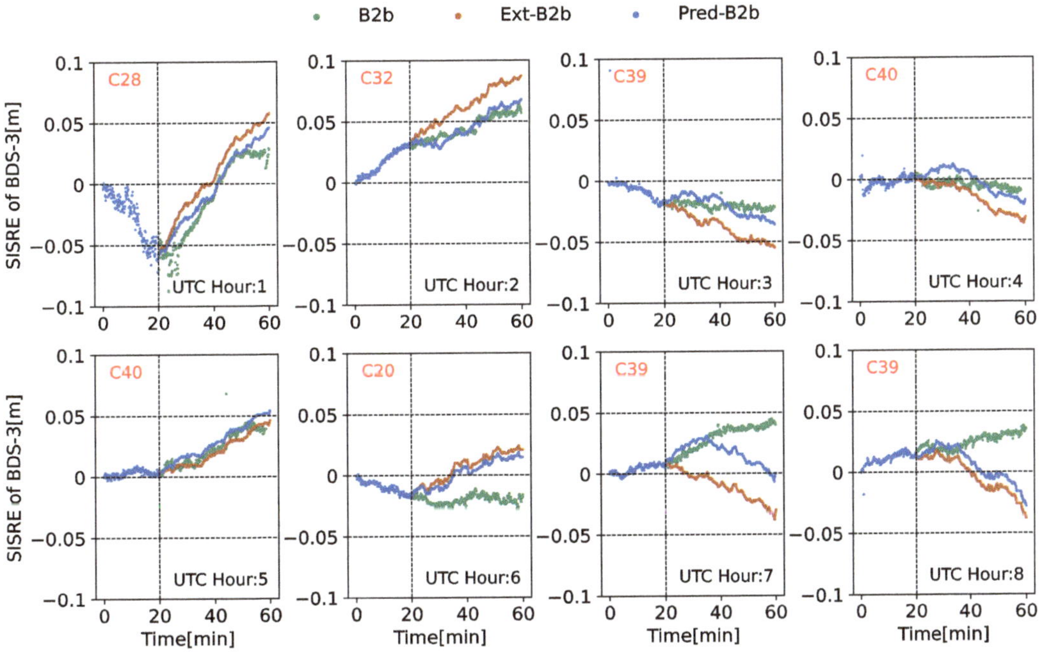

Figure 2. SISRE of BDS-3 for different interruption scenarios (spanning from 10 min to 40 min).

To further analyze the SISRE precision of the two prediction methods during signal interruptions, we processed hourly data over three days: the first twenty minutes of each hour received the B2b signal normally, followed by a prediction starting from the 20th minute. It is important to note that the generation of B2b products relies solely on reference stations in China and nearby regions, leading to a constant bias in the B2b clock corrections [30]. Consequently, we performed statistical analyses of the SISRE STD for the Ext and Pred prediction modes during interruption durations of 10, 20, 30, and 40 min (with B2b serving as a comparison without interruptions). Figures 4 and 5 illustrate the STD of the SISRE for each satellite, presented as bar charts under different conditions.

From Figure 4, it is evident that, under uninterrupted conditions, BDS-3 maintains a relatively concentrated SISRE over the short term, despite a gradual increase in the STD of the SISRE with longer observation times, remaining at around 0.01 m. In the case of a 10 min interruption, the STD of B2b, Ext-B2b, and Preb-B2b is comparable. However, Ext-B2b shows a lower precision relative to the other two methods. As the interruption duration increases, both Ext-B2b and Pred-B2b exhibit a noticeable decline in precision. For instance, with a 40 min interruption for the C26 satellite, the STD of the SISRE of the Ext-B2b exceeds 0.06 m, while our Pred-B2b model demonstrates superior precision compared to

Ext-B2b. In contrast, as shown in Figure 5, GPS shows inferior SISRE precision compared to BDS-3; under the 10 min interruption scenario, the STD of SISRE of the B2b model hovers at around 0.02 m. Nonetheless, the overall trend remains similar to that of BDS-3, where B2b offers the highest precision and Ext-B2b the lowest. Hence, for subsequent positioning tasks, the process noise of random walk assigned to individual satellites in Equation (12) can be adjusted based on their respective SISRE precision values. This approach facilitates the implementation of a positioning model that takes SISRE into account more effectively, ultimately enhancing the overall positioning performance.

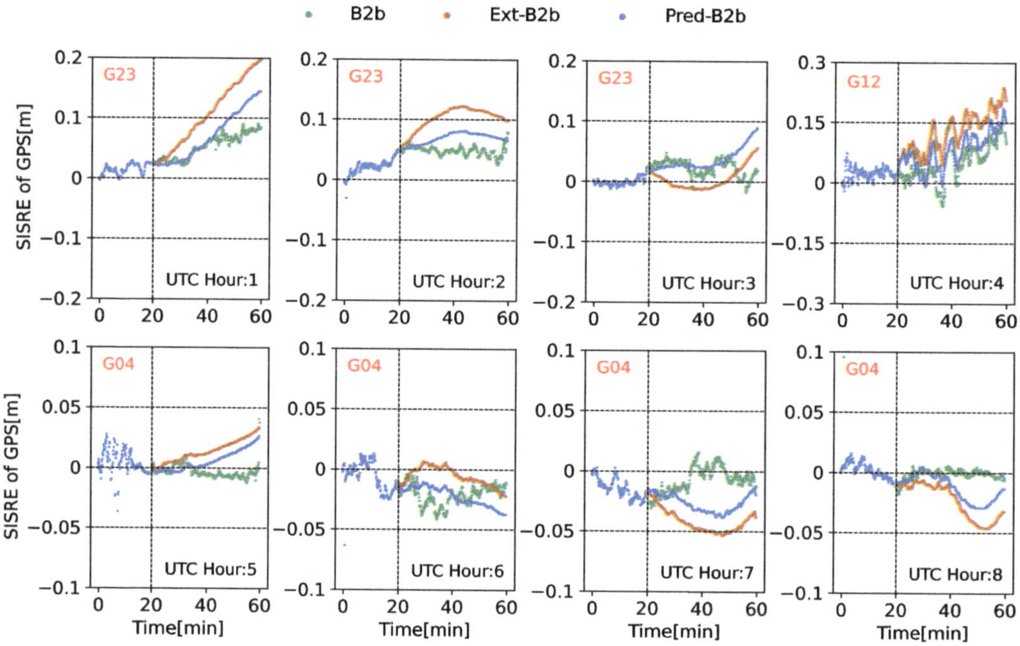

Figure 3. SISRE of GPS for different interruption scenarios (spanning from 10 min to 40 min).

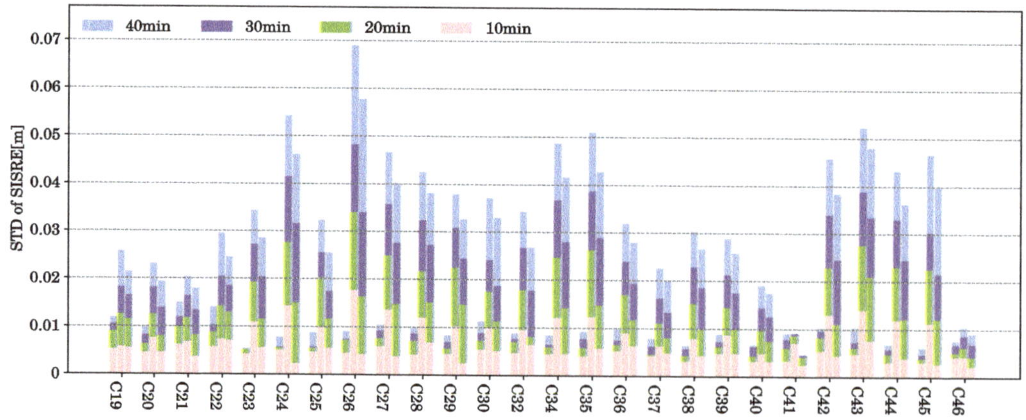

Figure 4. STD of SISRE for BDS-3 satellites under different interruption durations in non-Interrupted, Ext, and Pred Scenarios (Each label contains three bars representing the three methods: the first column represents B2b, the second column represents the Ext-B2b, and the third column represents the Pred-B2b).

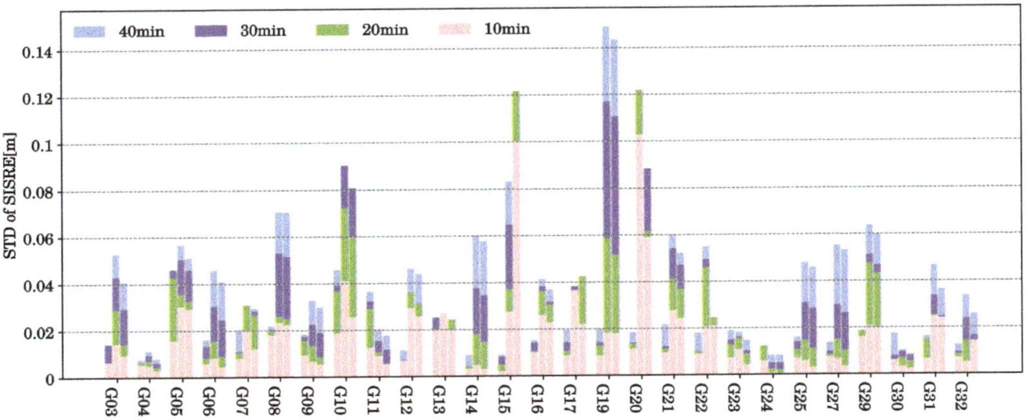

Figure 5. STD of SISRE for GPS satellites under different interruption durations in non-Interrupted, Ext, and Pred Scenarios (Each label contains three bars representing the three methods: the first column represents B2b, the second column represents the Ext-B2b, and the third column represents the Pred-B2b).

4. Analysis of Positioning Results

To determine the impact of predicted orbits and clock corrections on positioning, we conducted positioning experiments by collecting both simulated and real kinematic data over several days at different time intervals. The PPP data processing strategy is outlined in Table 1. Using the standard B2b-PPP as a reference, we compared the positioning performance of several models under various interruption durations: Ext-B2b-based PPP (EB2b-PPP), Pred-B2b-based PPP (PB2b-PPP), Ext-B2b-based PPP with SISRE adjustment (EB2bS-PPP), and Pred-B2b-based PPP with SISRE adjustment (PB2bS-PPP). It is worth noting that, for positioning after the interruption of the B2b signal, the most crucial factors encompass not only the accuracy of positioning throughout the entire interruption period but also the maximum positioning error and the stability of positioning upon the resumption of B2b signal reception. Consequently, in the subsequent analysis of the positioning results, we will delve into these three aspects separately.

Table 1. Summary of the PPP processing strategy.

Item	Processing Strategies
Observations	Dual frequency pseudo-range and carrier phase IF combinations
GNSS Signals	GPS (L1, L2), BDS-3 (B1I, B3I)
Satellite orbits and clock	(1) CNAV1, LNAV + B2b (B2b-PPP) (2) CNAV1, LNAV + Ext-B2b (EB2b-PPP) (3) CNAV1, LNAV + Pred-B2b (PB2b-PPP) (4) CNAV1, LNAV + Ext-B2b + SISRE (EB2bS-PPP) (5) CNAV1, LNAV + Pred-B2b + SISRE (PB2bS-PPP)
DCB	The DCB products from B2b [7]
SISRE	Estimated as random walk
Cutoff elevation	10°
Ambiguities	float

Table 1. *Cont.*

Item	Processing Strategies
Atmospheric delays	ZHD is corrected by the Saastamoinen model [35] and mapping function is from GMF [36]
Satellite PCO/PCV	igs14.atx [37]
Relativistic effect	Model corrected [38]
Phase windup	Model corrected [38]
Tide model	Solid tide and Ocean tide model corrected [38]
Receiver clock	Estimated as white noise, and the GPS and BDS-3 measurements are processed distinctively
Receiver position	White noise
Estimation	Kalman filter

4.1. Analysis of Positioning Preformance Under a Simulated Kinematic Experiment with Different Interruption Durations

The simulated kinematic data collection setup, as depicted in Figure 6, was implemented at the RTK (real time kinematic) base station (ZHD0) situated within the Jiulonghu campus of Southeast University. On DOY 120–121 in 2024, data were gathered from this station, which is furnished with a precision HI-TARGET AT-53501 antenna and a CHCNAV P5 receiver. To conduct a differentiated comparison that considers the characteristics of BRDC interruptions at the hour mark, we selected experimental data from 00:20 to 01:40 on DOY 120, and from 02:20 to 03:40 on DOY 121.

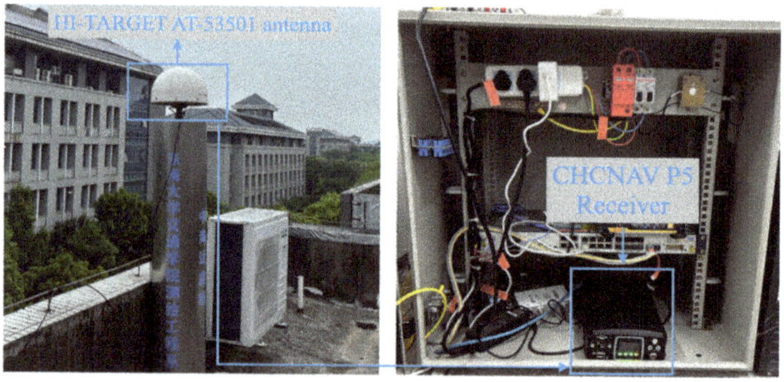

Figure 6. The ZHD0 station used for collecting static experiment data. The **left panel** displays the antenna used, while the **right panel** shows the receiver.

The observable number of satellites for the two data segments is shown in Figure 7. It can be observed that, except for the period from 1:15 to 1:37 on DOY 120, when the number of GPS satellites was relatively low, the number of satellites in the two segments was similar at other times. It is also evident that BDS has a higher number of observable satellites. However, since B2b only broadcasts corrections for BDS-3 satellites, the number of matched BDS-3 satellites is close to the observable number of GPS satellites. In comparison, after fusing GPS and BDS-3, the PDOP (position dilution of precision) can be maintained below 2. Therefore, in this test, we utilized the combined positioning results from BDS-3 and GPS.

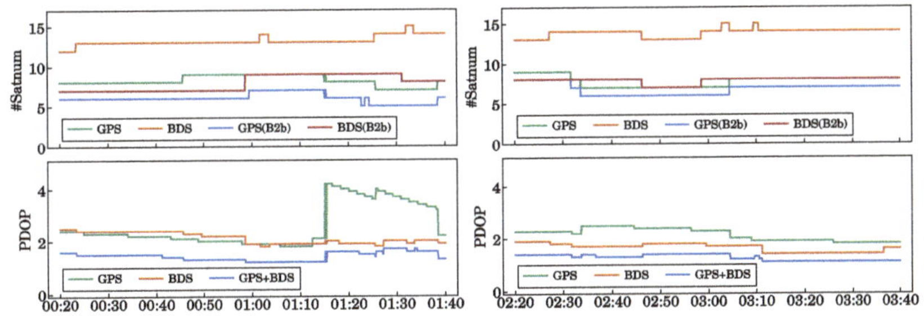

Figure 7. The effective satellites and PDOP for the ZHD0 on DOY 120 (**left**) and 121 (**right**).

On DOY 120, we initiated positioning at 0:20 and artificially interrupted the B2b data transmission at 0:40. For DOY 121, positioning commenced at 2:20, followed by an interruption in the B2b data transmission at 2:40. To validate the positioning accuracy under varying interruption durations, we conducted positioning tests with interruptions of 10 min, 20 min, 30 min, and 40 min for both sets of data. The root mean square (RMS) results of positioning in the E/N/U directions during the B2b signal interruption for DOY 120 and DOY 121 are presented in Tables 2 and 3 (for the sake of esthetic appeal, in the tables, we have set the unit of RMS to centimeters). Notably, B2b-PPP serves as the benchmark, representing the scenario where the B2b signal remains uninterrupted.

Table 2. Comparison of positioning accuracy among four prediction models under different interruption durations for DOY 120 (where B2b-PPP represents an uninterrupted B2b signal).

PPP Model	RMS of E/N/U Position Errors (cm)											
	10 min			20 min			30 min			40 min		
	E	N	U	E	N	U	E	N	U	E	N	U
B2b-ppp	1.2	1	1.7	1.3	1	1.9	1.2	1	2.8	1.2	1	2.7
EB2b-ppp	2.5	2.7	10.7	4.5	3.4	17.5	5.6	3.2	15.7	5.2	3.4	13.7
PB2b-ppp	1.7	1.4	1.5	2.5	2	5.1	3.5	2	6.1	3.9	2.6	7.1
EB2bS-ppp	1.7	1.7	3.9	3	3	2.3	1.8	4.1	3.5	3.3	2.6	11
PB2bS-ppp	1.3	1.1	1.5	1.5	1.5	2.5	1.5	1.9	2.8	1.9	2.1	3.9

Table 3. Comparison of positioning accuracy among four prediction models and B2b-PPP under different interruption durations for DOY 121.

PPP Model	RMS of E/N/U Position Errors (cm)											
	10 min			20 min			30 min			40 min		
	E	N	U	E	N	U	E	N	U	E	N	U
B2b-ppp	0.6	0.4	1.3	1	0.6	1.7	1.8	0.7	2.4	2	0.7	2.5
EB2b-ppp	1.6	1.6	2.8	4	2.5	9.1	5.4	2.6	4.3	5.1	2.9	4.7
PB2b-ppp	1.4	0.9	3.3	1.5	1.5	7.6	1.9	2	7.8	2	2.1	6.9
EB2bS-ppp	1.2	1.1	2.9	1.3	1	3.5	1.2	0.9	4.6	1.2	1	6.3
PB2bS-ppp	1.6	1	1.6	1.1	0.5	3.7	0.9	0.7	3.4	1	1.5	4.1

For DOY 120, it is evident that the EB2b-PPP method performs the worst, with positioning errors exceeding 0.025 m in the E/N directions and over 0.1 m in the U direction even with just a 10 min interruption in the B2b signal. From the results of EB2bS-PPP, it becomes clear that incorporating the SISRE significantly improves the positioning accuracy. This demonstrates the effectiveness of utilizing SISRE techniques to mitigate the impact of signal interruptions. Furthermore, our B2b signal prediction method shows a notable improvement for this dataset. The PB2b-PPP accuracy is comparable to that of EB2bS-PPP, indicating that the predicted B2b signals can effectively substitute for actual measurements during interruptions. Furthermore, PB2bS-PPP emerges as the best-performing method, achieving a positioning accuracy comparable to B2b-PPP even when the B2b signal is interrupted for 40 min. This highlights the combined benefits of B2b corrections prediction and SISRE enhancements in maintaining high positioning accuracy under challenging conditions.

For DOY 121, it can be observed that EB2b-PPP demonstrates a relatively good positioning performance within this dataset. However, as shown in Table 3, under various interruption conditions, PB2bS-PPP outperforms the other positioning methods. Overall, EB2b-PPP exhibits the lowest positioning accuracy among the models compared. EB2bS-PPP, which incorporates enhancement techniques, slightly surpasses PB2b-PPP in positioning precision. When focusing on horizontal accuracy (E and N directions), EB2bS-PPP and PB2bS-PPP show similar performances. In contrast, for vertical accuracy (U), PB2bS-PPP clearly stands out as the most accurate model.

It is worth noting that both Tables 2 and 3 show cases where longer interruptions result in a better positioning accuracy than shorter interruptions. For example, in Table 2, for EB2bS-PPP, the positioning accuracy in the U direction with a 20 min interruption is better than that with a 10 min interruption. In Table 3, for PB2bS-PPP, the positioning accuracy in both the E and N directions is superior with a 30 min interruption compared to a 10 min interruption. This is because, in practical positioning, B2b corrections do not change monotonically over time. Therefore, the positioning results do not necessarily increase or decrease monotonically with the predicted corrections, leading to the phenomenon where longer interruptions result in better positioning accuracy than shorter ones.

To further illustrate this, we have plotted the positioning errors in the E/N/U directions for five positioning models across different interruption durations in Figure 8. In Figure 8, the black vertical lines represent the moments when B2b signal interruption began, while the blue vertical lines indicate the resumption of B2b signal reception. From Figure 8, we can clearly observe the positioning fluctuations, and draw conclusions like those in Tables 2 and 3, namely that PB2bS-PPP aligns most favorably with B2b-PPP. Furthermore, from Figure 8, we can see that the maximum positioning error is one of the key indicators for evaluating positioning performance during B2b signal interruptions. Therefore, Figure 9 showcases the maximum positioning errors in the E/N/U directions across four distinct time intervals.

Figure 9 reveals that, in terms of horizontal positioning accuracy, PB2bS-PPP exhibits maximum positioning errors below 0.04 m, while PB2b-PPP and EB2bS-PPP both maintain maximum errors below 0.06 m. In contrast, EB2b-PPP's maximum positioning deviation surpasses 0.1 m. Regarding the vertical dimension, EB2b-PPP, EB2bS-PPP, and PB2b-PPP record maximum positioning deviations exceeding 0.1 m. Notably, EB2b-PPP stands out with deviations that even exceed 0.3 m, underscoring its significant shortcoming in this regard. Conversely, PB2bS-PPP demonstrates a more contained deviation, with all values falling below 0.1 m, highlighting its relative superiority in vertical positioning.

From Figure 8, we can also gain further insights into the positioning stability of various models following the restoration of B2b signals. For the four prediction models applied to the two sets of data, positioning remains relatively stable both before and immediately after the 10 min interruption of B2b signals. However, as the duration of the interruption increases, EB2b-PPP struggles to converge to a positioning accuracy comparable to that of B2b-PPP upon signal restoration. Additionally, EB2bS-PPP and PB2b-PPP also exhibit instances of non-convergence, such as in the case of EB2bS-PPP for the E direction on

DOY 120 after a 40 min interruption, and PB2b-PPP for the U direction on DOY 121 after a 20 min interruption, with both failing to achieve satisfactory results. In contrast, PB2bS-PPP consistently demonstrates satisfactory stability before and after signal restoration, underscoring its importance for practical dynamic positioning applications. This robustness is clearly crucial for ensuring reliable and accurate positioning in dynamic environments.

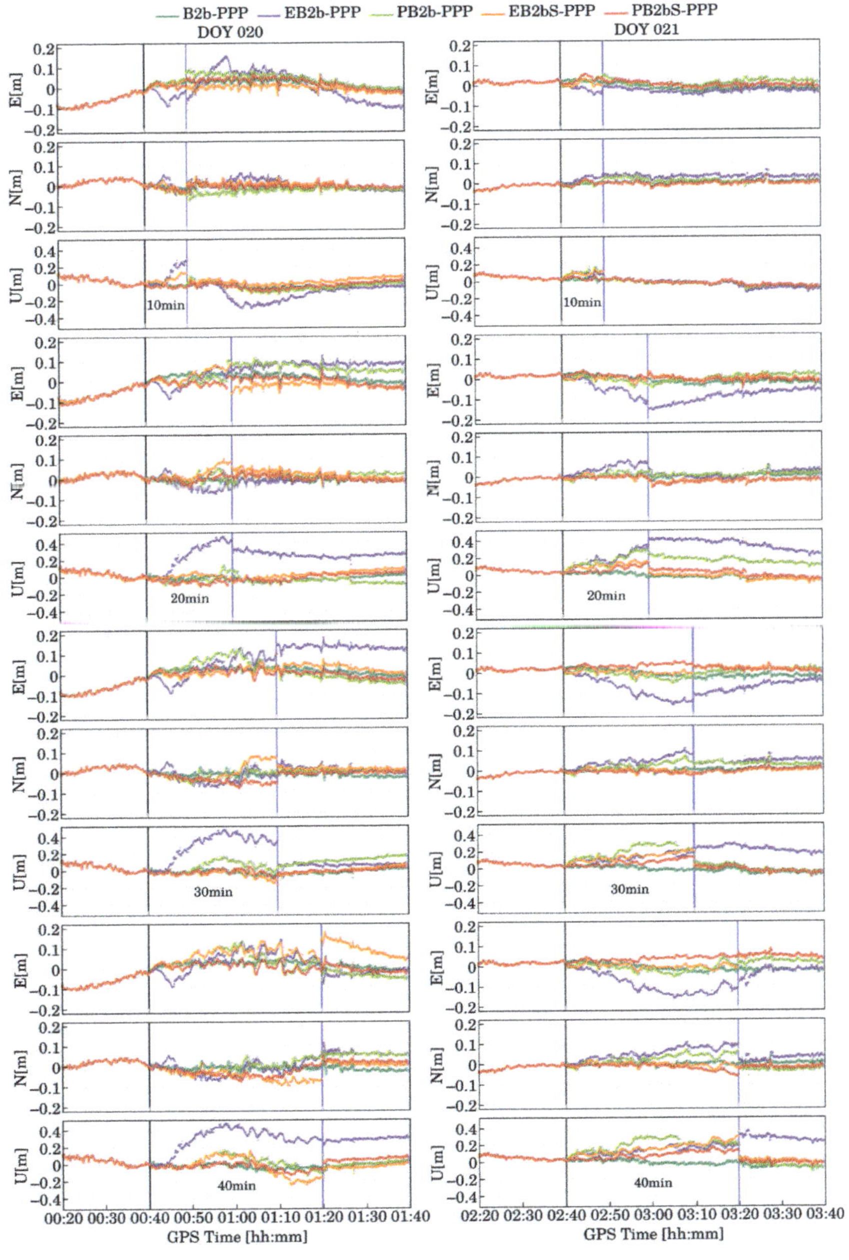

Figure 8. The positioning errors in the E/N/U directions for the B2b-PPP, EB2b-PPP, PB2b-PPP, EB2bS-PPP, and PB2bS-PPP under signal interruptions of 10 min, 20 min, 30 min, and 40 min on DOY 120 and 121.

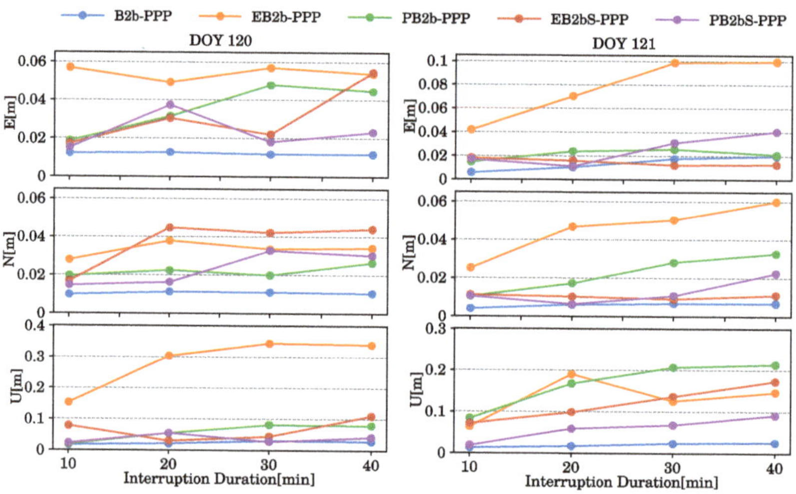

Figure 9. The maximum positioning errors in the E/N/U directions for the B2b-PPP, EB2b-PPP, PB2b-PPP, EB2bS-PPP, and PB2bS-PPP under signal interruptions of 10 min, 20 min, 30 min, and 40 min on DOY 120 and 121.

4.2. Analysis of Positioning Performance Under Kinematic Experiment with Different Interruption Durations

To demonstrate the kinematic positioning accuracy of B2b-PPP prediction and positioning model under B2b signal interruption conditions, we conducted practical kinematic positioning tests in Nanjing on DOY 177, 2024. As illustrated in the left panel of Figure 10, a CHCNAV I90 receiver was mounted on the roof of a vehicle to capture kinematic BDS-3 and GPS observation data during the test. The CGI-430 GNSS/INS integrated navigation device from CHCNAV served as the benchmark for kinematic positioning, and the base station ZHD0 in Figure 6 provided the RTK reference for CGI-430. Concurrently, we employed the board depicted in Figure 2 to receive B2b corrections in real-time for B2b-PPP. As evidenced by the test trajectory in Figure 10 (right), there were no GNSS interruptions recorded throughout our entire test route, and the maximum linear distance between the test route and the ZHD0 station was 3.7 km. Under such circumstances, the nominal accuracy of the CGI-430 can achieve 0.02 m in the horizontal and 0.03 m in the vertical direction, providing a precise and genuine positioning accuracy reference for evaluating our B2b-PPP approach. Notably, when outputting the dynamic positioning results from the CGI-430, we conducted lever arm corrections to align the positioning outcomes with those from the I90 receiver.

The entire positioning test lasted for 60 min, spanning from GPS time 06:40 to 07:40. Figure 11 presents the number of satellites received during dynamic testing, the number of satellites matched in B2b, and the corresponding PDOP values. During the period from 6:40 to 6:50, we utilized B2b corrections for normal positioning. Starting from 6:50, we artificially interrupted the B2b signals and conducted a comparative analysis of positioning accuracy among the four proposed positioning models under B2b signal interruption scenarios. The statistical results of positioning accuracy for interruptions lasting 10 min, 20 min, 30 min, and 40 min are presented in Table 4. Furthermore, to provide a more intuitive comparison of the positioning performance of the four models, Figure 12 showcases the positioning accuracy of the five models under a 40 min interruption scenario, as well as the convergence situation after B2b signal restoration at 7:30. This visualization enables a clear assessment of how each model copes with extended signal outages and their subsequent recovery capabilities.

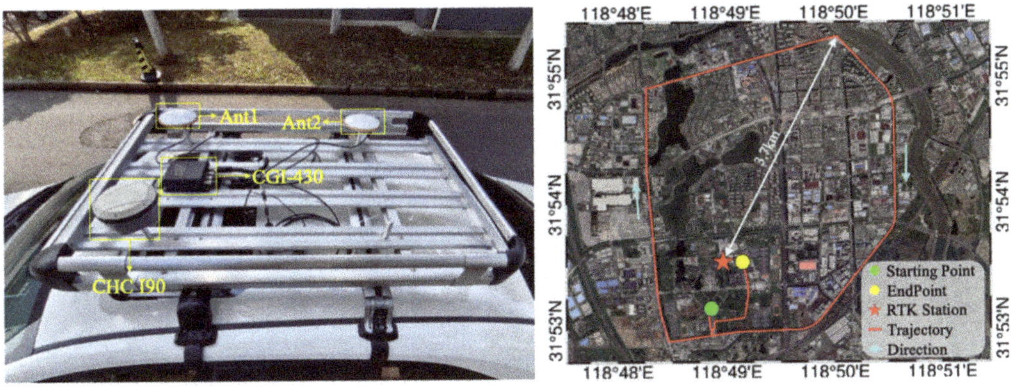

Figure 10. Dynamic vehicle-mounted positioning experimental setup (**left**) and trajectory of the measurement campaign (**right**).

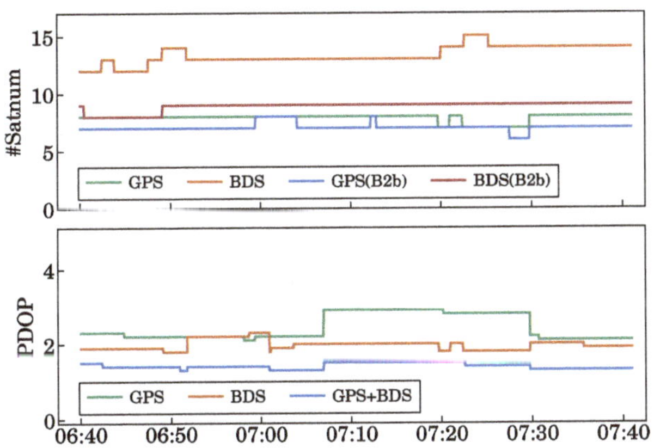

Figure 11. The number of effective satellites and PDOP for the kinematic experiment.

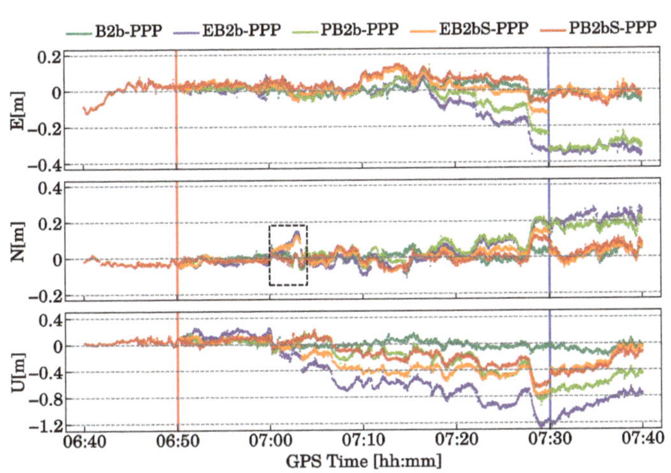

Figure 12. Positioning errors in the E/N/U directions for various PPP models (B2b-PPP, EB2b-PPP, PB2b-PPP, EB2bS-PPP, and PB2bS-PPP) during a 40 min B2b signal interruption in the kinematic experiment.

Table 4. Comparison of positioning accuracy among four prediction models and B2b-PPP under different interruption durations for the dynamic scenario.

PPP Model	RMS of E/N/U Position Errors (cm)											
	10 min			20 min			30 min			40 min		
	E	N	U	E	N	U	E	N	U	E	N	U
B2b-ppp	2.2	6.3	18.8	3.7	6.4	15.1	3.3	5.9	15.5	3.3	6.1	14.1
EB2b-ppp	2.7	8.4	26.9	3.0	8.0	29.7	4.2	8.2	38.4	11.8	7.9	51.7
PB2b-ppp	1.6	7.2	16.8	3.4	6.4	16.0	3.9	6.3	14.0	8.3	6.4	22.6
EB2bS-ppp	1.8	7.7	19.5	3.1	6.8	17.5	3.8	7.4	22.3	4.9	6.9	27.1
PB2bS-ppp	2.2	7.4	18.1	2.2	7.0	17.4	5.6	9.8	14.6	4.4	7.1	18.9

As is evident from Table 4, under uninterrupted B2b signal conditions and in dynamic scenarios, the positioning accuracy of B2b-PPP in the E/N direction is approximately 0.04 m to 0.06 m, while in the U direction, it remains around 0.15 m. Similar to simulated dynamic positioning, EB2b-PPP exhibits the poorest positioning performance, with a U-direction positioning RMS of up to 0.269 m after just 10 min of interruption. Although PB2bS-PPP also experiences an increase in positioning error with longer interruption durations, it maintains a positioning RMS below 0.1 m in the E/N direction even after 40 min. Compared to the two direct prediction models mentioned earlier, EB2bS-PPP and PB2bS-PPP show a similar positioning accuracy in the horizontal direction, but PB2bS-PPP outperforms in the U direction, with an RMS below 0.2 m.

Additionally, Figure 12 shows that when interruptions last for 40 min, the maximum positioning errors in the two SISRE-corrected models are significantly lower than those of models that do not account for SISRE. Specifically, PB2bS-PPP has a maximum E/N positioning error of less than 0.1 m, and in the U direction, it maintains a maximum error of less than 0.4 m for interruptions not exceeding 37 min. Following signal restoration, both EB2bS-PPP and PB2bS-PPP converge more rapidly or align directly with the uninterrupted B2b signal scenario, highlighting the improvement in predicting B2b corrections when SISRE is considered. Notably, as indicated by the black boxes in Figure 12, directly using B2b extension results can lead to abnormal and sudden changes in positioning. Therefore, in actual B2b signal interruption scenarios, we recommend utilizing PB2bS-PPP to ensure high-precision positioning over extended periods.

5. Conclusions

Due to the "south wall effect", signal matching priority, and IOD matching errors, B2b signals in dynamic positioning are susceptible to obstruction by buildings, mountains, and other structures, resulting in positioning interruptions. Existing B2b signal prediction models inevitably suffer from discrepancies between the predicted and actual signals. As the prediction time increases, positioning accuracy declines rapidly, and when the B2b signal is restored, positioning jumps can occur due to the deviation between the predicted and real signals. In this paper, we introduce SISRE parameters to absorb residual errors from the predicted signals, and develop observation and stochastic models that incorporate SISRE, aiming to achieve continuous, seamless, high-precision positioning based on B2b.

First, we analyzed the SISRE of three B2b products, B2b, Ext-B2b, and Pred-B2b, during interruptions of 10, 20, 30, and 40 min. The results revealed that the B2b SISRE for BDS-3 generally remains below 0.1 m, while the SISRE for GPS occasionally exceeds 0.15 m. Additionally, the Pred-B2b showed better alignment with the actual B2b signal during signal interruptions. Furthermore, a three-day statistical analysis indicated that the STD of the B2b SISRE for BDS-3 is primarily below 0.01 m, whereas for GPS, it is mostly under

0.02 m. As the interruption duration increases, the SISRE for both Ext-B2b and Pred-B2b gradually declines, with Pred exhibiting superior performance metrics.

Simulated dynamic positioning experiments revealed that the EB2b-PPP performance is the weakest, with positioning errors in the U direction exceeding 0.1 m during a 10 min interruption. In contrast, the performances of EB2bS-PPP and PB2b-PPP are comparable, though the PB2bS-PPP model demonstrated the best overall positioning performance. After a 40 min interruption, the positioning RMS of PB2bS-PPP in the E/N/U directions remained below 0.05 m, with the maximum positioning deviation not exceeding 0.1 m. Furthermore, upon re-acquisition of the B2b signal, the PB2bS-PPP model exhibited a smooth transition in positioning.

Although the positioning accuracy of the real dynamic positioning experiments is lower than that of the simulated dynamic positioning experiments, their performance results are similar. During a 40 min interruption, the PB2bS-PPP model exhibited an RMS in the E/N directions of less than 0.1 m, with a maximum positioning deviation also under 0.1 m. The RMS in the U direction is also below 0.2 m, and the maximum positioning deviation is better than 0.4 m within a 37 min interruption. Furthermore, upon restoration of the B2b signal, stable positioning was maintained. Therefore, the PB2bS-PPP model proposed in this paper effectively ensures continuous and stable positioning based on B2b signals.

Author Contributions: Conceptualization, R.S. and Z.X.; methodology, R.S.; software, Z.X. validation, Q.L.; formal analysis, R.S.; investigation, R.S. and Z.X.; resources, R.S. and W.G.; data curation, Z.X. and R.S.; writing original draft preparation, R.S.; writing—review and editing, C.G., R.S., X.M. and W.G.; visualization, Z.X., R.S. and Q.L.; supervision, C.G. and X.M.; project administration R.S.; funding acquisition R.S. All authors have read and agreed to the published version of the manuscript.

Funding: This work was supported by the National Natural Science Foundation of China (No. 42404024) and Jiangsu Funding Program for Excellent Postdoctoral Talent (No. 2022ZB113).

Data Availability Statement: The data that support the findings of this study are available on request from the corresponding author, upon reasonable request.

Acknowledgments: The authors gratefully acknowledge the valuable comments from the anonymous reviewers.

Conflicts of Interest: The authors declare no conflicts of interest.

References

1. Malys, S.; Jensen, P.A. Geodetic point positioning with GPS carrier beat phase data from the CASA UNO experiment. *Geophys. Res. Lett.* **1990**, *17*, 651–654. [CrossRef]
2. Zumberge, J.; Heflin, M.; Jefferson, D.; Watkins, M.; Webb, F.H. Precise point positioning for the efficient and robust analysis of GPS data from large networks. *J. Geophys. Res. Solid. Earth* **1997**, *102*, 50055017. [CrossRef]
3. An, X.; Meng, X.; Jiang, W. Multi-constellation GNSS precise point positioning with multi-frequency raw observations and dual-frequency observations of ionospheric-free linear combination. *Satell. Navig.* **2020**, *1*, 7. [CrossRef]
4. Zhao, Q.; Pan, S.; Gao, C.; Gao, W.; Xia, Y. BDS/GPS/LEO triple-frequency uncombined precise point positioning and its performance in harsh environments. *Measurement* **2020**, *151*, 107216. [CrossRef]
5. Xiao, P.; Sun, F.; Wang, K.; Xiao, K.; Shang, X.; Liu, J. Positioning performance analysis of real-time BDS-3 PPP-B2b/INS tightly coupled integration in urban environments. *Adv. Space Res.* **2023**, *72*, 4008–4020. [CrossRef]
6. Shang, R.; Gao, C.; Meng, X.; Xu, Z.; Liu, Q.; Chen, Q.; An, X. Lightweight seamless lane-level positioning with the integration of BDS-3 PPP-B2b and VINS: A case study using smartphone. *GPS Solut.* **2024**, *28*, 185. [CrossRef]
7. CSNO. BeiDou Navigation Satellite System Signal in Space Interface Control Document Precise Point Positioning Service Signal PPP-B2b (Version 1.0). 2020. Available online: http://www.beidou.gov.cn/xt/gfxz/202008/P020200803362062482940.pdf (accessed on 5 January 2025).
8. Liu, Y.; Yang, C.; Zhang, M. Comprehensive Analyses of PPP-B2b Performance in China and Surrounding Areas. *Remote Sens.* **2022**, *14*, 643. [CrossRef]

9. Lu, X.; Chen, L.; Shen, N.; Wang, L.; Jiao, Z.; Chen, R. Decoding PPP corrections from BDS B2b signals using a software-defined receiver: An initial performance evaluation. *IEEE Sens. J.* **2020**, *21*, 7871–7883. [CrossRef]
10. Zhang, W.; Lou, Y.; Song, W.; Sun, W.; Zou, X.; Gong, X. Initial assessment of BDS-3 precise point positioning service on GEO B2b signal. *Adv.Space Res.* **2022**, *69*, 690–700. [CrossRef]
11. Zhou, H.; Wang, L.; Fu, W.; Han, Y.; Li, T.; Li, W.; Chen, R. Real-time single-frequency precise point positioning using BDS-3 PPP-B2b corrections. *Measurement* **2022**, *205*, 112178. [CrossRef]
12. Ge, Y.; Cao, X.; Lyu, D.; He, Z.; Xiao, G.; Shen, F. An investigation of PPP time transfer via BDS-3 PPP-B2b service. *GPS Solut.* **2023**, *27*, 61. [CrossRef]
13. Ren, Z.; Gong, H.; Peng, J.; Tang, C.; Huang, X.; Sun, G. Performance assessment of real-time precise point positioning using BDS PPP-B2b service signal. *Adv. Space Res.* **2021**, *68*, 3242–3254. [CrossRef]
14. Wu, P.; Lou, Y.; Zhang, W.; Jan, D.; He, H.; Chai, J.; Ouyang, Y.; Zhang, Z.; Zou, X. Evaluation of real-time kinematic positioning performance of the BDS-3 PPP service on B2b signal. *GPS Solut.* **2023**, *27*, 192. [CrossRef]
15. Xu, Y.; Yang, Y.; Li, J. Performance evaluation of BDS-3 PPP-B2b precise point positioning service. *GPS Solut.* **2021**, *25*, 142. [CrossRef]
16. European Union. Galileo High Accuracy Service Signal-InSpace Interface Control Document (HAS SIS ICD) Issue 1.0. 2022. Available online: https://www.gsc-europa.eu/sites/default/files/sites/all/files/Galileo_HAS_SIS_ICD_v1.0.pdf (accessed on 5 January 2025).
17. European Union. Galileo High Accuracy Service Service Definition Document (HAS SDD) Issue 1.0. 2023. Available online: https://www.gsc-europa.eu/sites/default/files/sites/all/files/Galileo-HAS-SDD_v1.0.pdf (accessed on 5 January 2025).
18. Yang, Y.; Liu, L.; Li, J.; Yang, Y.; Zhang, T.; Mao, Y.; Sun, B.; Ren, X. Featured Services and Performance of BDS-3. *Sci. Bull.* **2021**, *66*, 2135–2143. [CrossRef]
19. Yang, Y.; Ding, Q.; Gao, W.; Li, J.; Xu, Y.; Sun, B. Principle and performance of BDSBAS and PPP-B2b of BDS-3. *Satell. Navig.* **2022**, *3*, 5. [CrossRef]
20. Yang, Y.; Mao, Y.; Sun, B. Basic performance and future developments of BeiDou global navigation satellite system. *Satell. Navig.* **2020**, *1*, 1. [CrossRef]
21. Tao, J.; Liu, J.; Hu, Z.; Zhao, Q.; Chen, G.; Ju, B. Initial Assessment of the BDS-3 PPP-B2b RTS compared with the CNES RTS. *GPS Solut.* **2021**, *25*, 131. [CrossRef]
22. Xu, X.; Nie, Z.; Wang, Z.; Zhang, Y.; Dong, L. An Improved BDS-3 PPP-B2b Positioning Approach by Estimating Signal in Space Range Errors. *GPS Solut.* **2023**, *27*, 110. [CrossRef]
23. Xu, Z.; Shang, R.; Gao, C.; Gao, W.; Liu, Q.; Long, F.; Xu, D. Temporal Characteristics Based Outlier Detection and Prediction Methods for PPP-B2b Orbit and Clock Corrections. *Remote Sens.* **2024**, *16*, 2337. [CrossRef]
24. Ge, H.; Meng, G.; Li, B. Zero-reconvergence PPP for real-time low-earth satellite orbit determination in case of data interruption. *IEEE J. Sel. Top. Appl. Earth Obs. Remote Sens.* **2024**, *17*, 4705–4715. [CrossRef]
25. Chen, Y.; Wang, X.; Zhou, K.; Zhang, J.; Qiu, C.; Li, H.; Xin, S. Real-Time Precise Point Positioning during Outages of the PPP-B2b Service. *Remote Sens.* **2023**, *15*, 784. [CrossRef]
26. Montenbruck, O.; Steigenberger, P.; Hauschild, A. Multi-GNSS signal-in-space range error assessment—Methodology and results. *Adv. Space Res.* **2018**, *61*, 3020–3038. [CrossRef]
27. Carlin, L.; Hauschild, A.; Montenbruck, O. Precise point positioning with GPS and Galileo broadcast ephemerides. *GPS Solut.* **2021**, *25*, 77. [CrossRef]
28. Chen, G.; Wei, N.; Li, M.; Zhao, Q.; Zhang, J. BDS-3 and GPS/Galileo integrated PPP using broadcast ephemerides. *GPS Solut.* **2022**, *26*, 129. [CrossRef]
29. Tang, C.; Hu, X.; Chen, J.; Liu, L.; Zhou, S.; Guo, R.; Li, X.; He, F.; Liu, J.; Yang, J. Orbit determination, clock estimation and performance evaluation of BDS-3 PPP-B2b service. *J. Geod.* **2022**, *96*, 60. [CrossRef]
30. Sun, S.; Wang, M.; Liu, C.; Meng, X.; Ji, R. Long-term performance analysis of BDS-3 precise point positioning (PPP-B2b) service. *GPS Solut.* **2023**, *27*, 69. [CrossRef]
31. Yao, Y.; He, Y.; Yi, W.; Song, W.; Cao, C.; Chen, M. Method for evaluating real-time GNSS satellite clock offset products. *GPS Solut.* **2017**, *21*, 1417–1425. [CrossRef]
32. Lou, Y.; Zheng, F.; Gu, S.; Wang, C.; Guo, H.; Feng, Y. Multi-GNSS precise point positioning with raw single-frequency and dual-frequency measurement models. *GPS Solut.* **2016**, *20*, 849–862. [CrossRef]
33. Zhou, F.; Dong, D.; Li, P.; Li, X.; Schuh, H. Influence of stochastic modeling for inter-system biases on multi-GNSS undifferenced and uncombined precise point positioning. *GPS Solut.* **2019**, *23*, 59. [CrossRef]
34. Takasu, T.; Yasuda, A. Development of the low-cost RTK-GPS receiver with an open source program package RTKLIB. In Proceedings of the International Symposium on GPS/GNSS. International Convention Center, Jeju, Republic of Korea, 12–17 July 2009; Volume 1. Available online: https://gpspp.sakura.ne.jp/paper2005/isgps_2009_rklib.pdf (accessed on 5 January 2025).

35. Saastamoinen, J. Atmospheric correction for the troposphere and stratosphere in radio ranging satellites. The use of artificial satellites for geodesy, American Geophysics Union. *Geophys. Monogr. Ser.* **1972**, *15*, 247–251. [CrossRef]
36. Böhm, J.; Niell, A.; Tregoning, P.; Schuh, H. Global Mapping Function (GMF): A new empirical mapping function based on numerical weather model data. *Geophys. Res. Lett.* **2006**, *33*, L07304. [CrossRef]
37. Rebischung, P.; Schmid, R. IGS14/igs14.atx: A new framework for the IGS products. In Proceedings of the AGU Fall Meeting, San Francisco, CA, USA, 12–16 December 2016.
38. Kouba, J. A Guide to Using International GNSS Service (IGS) Products. 2009. Available online: https://files.igs.org/pub/resource/pubs/UsingIGSProductsVer21_cor.pdf (accessed on 5 February 2025).

remote sensing

MDPI

Article

Localization of Multiple GNSS Interference Sources Based on Target Detection in C/N_0 Distribution Maps

Qidong Chen, Rui Liu *, Qiuzhen Yan, Yue Xu, Yang Liu, Xiao Huang and Ying Zhang

China Research Institute of Radiowave Propagation, Qingdao 266000, China; chenqd@crirp.ac.cn (Q.C.); yanqz@crirp.ac.cn (Q.Y.); 13045190872@163.com (Y.X.); leo123yang@163.com (Y.L.); huangx@crirp.ac.cn (X.H.); wfzywfzy@163.com (Y.Z.)

* Correspondence: liur@crirp.ac.cn

Abstract: The localization of multiple interference sources in Global Navigation Satellite Systems (GNSS) can be achieved using carrier-to-noise ratio (C/N_0) information provided by GNSS receivers, such as those embedded in smartphones. However, in increasingly prevalent complex scenarios—such as the coexistence of multiple directional interferences, increased diversity and density of GNSS interference, and the presence of multiple low-power interference sources—conventional localization methods often fail to provide reliable results, thereby limiting their applicability in real-world environments. This paper presents a multi-interference sources localization method using object detection in GNSS C/N_0 distribution maps. The proposed method first exploits the similarity between C/N_0 data reported by GNSS receivers and image grayscale values to construct C/N_0 distribution maps, thereby transforming the problem of multi-source GNSS interference localization into an object detection and localization task based on image processing techniques. Subsequently, an Oriented Squeeze-and-Excitation-based Faster Region-based Convolutional Neural Network (OSF-RCNN) framework is proposed to process the C/N_0 distribution maps. Building upon the Faster R-CNN framework, the proposed method integrates an Oriented RPN (Region Proposal Network) to regress the orientation angles of directional antennas, effectively addressing their rotational characteristics. Additionally, the Squeeze-and-Excitation (SE) mechanism and the Feature Pyramid Network (FPN) are integrated at key stages of the network to improve sensitivity to small targets, thereby enhancing detection and localization performance for low-power interference sources. The simulation results verify the effectiveness of the proposed method in accurately localizing multiple interference sources under the increasingly prevalent complex scenarios described above.

Keywords: GNSS interference; interference localization; target detection; C/N_0 distribution map

1. Introduction

GNSS generally refer to various global satellite navigation systems and their augmentation systems, including the U.S. Global Positioning System (GPS), China's BeiDou, Russia's Global Navigation Satellite System (GLONASS), and Europe's Galileo, among others [1,2]. GNSS is the preferred technology for most location- and time-dependent applications [3], supporting critical infrastructure closely tied to national operations and people's livelihoods [4]. The spread-spectrum modulation techniques employed in satellite navigation provide GNSS signals with a certain level of anti-jamming capability [5]. However, the low received power of GNSS signals makes the system vulnerable to electromagnetic environments or interfering signals [6,7].

Currently, GNSS systems face increasingly severe interference threats, with interference incidents occurring more frequently. Therefore, rapid and effective interference source localization has become a crucial means of ensuring GNSS application security [8,9]. To enhance the effectiveness of interference while improving concealment, multiple interference sources are often employed cooperatively to target GNSS signals. Moreover, low-cost jammers are now readily available on the commercial market [10], making multi-source GNSS interference an even greater threat than single-source interference [11]. With the widespread adoption of null-steering antennas, an array antenna composed of M elements can suppress M-1 interference sources [12]. This has led to the evolution of interference methods from single-source to multi-source cooperative jamming [13]. Such cooperative interference typically employs distributed low-power directional jamming, leading to increasingly common yet challenging scenarios, including multiple directional interference sources, growing types and quantities of GNSS interference, and multiple weak interference signals. These developments pose significant challenges to existing GNSS interference localization methods. This paper focuses on addressing the multi-interference source localization problem in these challenging scenarios.

1.1. Related Work

Common GNSS interference source localization methods include Direction of Arrival (DOA)-based [14], Time Difference of Arrival (TDOA)-based [15,16] and Received Signal Strength (RSS)-based [17,18] approaches. Among these, DOA-based localization methods are limited by the number of antenna array elements, and the commonly used cross-localization technique suffers from positioning ambiguity when multiple interference sources coexist [19]. TDOA-based methods, however, face challenges in accurately extracting time differences between various interference sources and receivers due to the difficulty in separating multiple interfering signals. Additionally, commercial off-the-shelf (COTS) receivers typically cannot provide the raw data samples required for TDOA estimation [20].

Due to their low implementation complexity, RSS-based methods have been widely adopted for GNSS interference source localization [21]. The concept of Jamming 911 (J911) was first proposed by Logan Scott at the 2011 Institute of Navigation (ION) GNSS conference. It introduced a crowdsourcing-based approach for interference detection and localization, leveraging carrier-to-noise ratio (C/N_0) measurements reported by smartphones to identify and locate GNSS jammers [22]. Smartphones enjoy near-ubiquitous global penetration. For instance, China's Ministry of Industry and Information Technology (MIIT) reported in its 2023 Communications Industry Statistical Bulletin (published in January 2024) that the country hosts 11.62 million mobile communication base stations and 1.727 billion mobile phone subscribers [23]. However, the developer of the J911 system did not specify how to efficiently process such massive-scale GNSS receiver data, nor did they provide an effective solution for multi-interference source localization [24]. Reference [25] proposed a centroid-based localization method for multiple GNSS interference sources using clustering and centroid convergence algorithms on C/N_0 data. While this method achieves multi-source localization, its performance degrades as the number and variety of interference sources increase. Moreover, it fails to address scenarios involving directional GNSS jammers or low-power interference sources.

1.2. Our Contribution

The paper begins by establishing a system and signal model tailored to the GNSS interference source localization scenario under study. Building upon this foundation, it proposes a multi-interference sources localization method based on object detection in GNSS C/N_0 distribution maps. The method innovatively constructs C/N_0 distribution maps by exploiting the similarity between GNSS receiver-reported data and image grayscale information. This transforms the GNSS multi-interference sources localization problem into an image processing-based object detection and localization task. To address key challenges in detecting and localizing target interference sources—including estimation of the main lobe direction for directional antennas, multi-target feature recognition and classification, and the detection/localization of low-power GNSS interference sources—the paper proposes an OSF-RCNN framework, an enhanced variant of Faster R-CNN, for processing C/N_0 distribution maps. The OSF-RCNN framework incorporates an Oriented RPN to regress the orientation angles of directional antennas, thereby enhancing both detection accuracy and localization precision for directional interference sources. The framework employs an FPN to achieve multi-scale feature fusion, which improves the method's adaptability to various interference types and different transmission power levels. Additionally, SE blocks are strategically integrated at critical network nodes to intensify focus on low-power interference targets, consequently boosting detection precision for the small low-power targets.

The study utilized the established system and signal model to construct a dedicated dataset, followed by comprehensive simulation tests to evaluate the proposed method's performance. The experimental validation specifically assessed the method's localization capabilities under the following challenging conditions: (1) directional interference, (2) multiple interference types, and (3) weak interference scenarios. The results demonstrate that the proposed OSF-RCNN method achieves effective detection and accurate localization of multiple GNSS interference sources across all these challenging operational conditions.

The remainder of this paper is organized as follows: Section 2 introduces the system and signal model of the scenarios discussed. In addition, this section describes the specific steps of the proposed multi-interference sources localization method based on target detection in C/N_0 distribution maps. It also presents the construction method for C/N_0 distribution maps, explains the dataset generation process, and analyzes the remaining challenges in detecting and localizing target interference sources in complex scenarios. In response to these challenges, the OSF-RCNN framework is designed and proposed. Section 3 conducts simulation tests to evaluate the method's performance and analyzes the test results. Section 4 presents the conclusions of this paper.

2. Materials and Methods

The paper proposes a localization method for multiple GNSS interference sources based on target detection in C/N_0 distribution maps. The method begins by constructing C/N_0 distribution maps, utilizing the similarity between GNSS receiver C/N_0 measurements and image grayscale features. Then, an OSF-RCNN framework is introduced to process the C/N_0 distribution map, enabling effective GNSS interference localization in challenging scenarios such as the presence of multiple directional interference sources, diverse and numerous interference types, and multiple weak interference signals. A detailed description of the proposed method is provided in this chapter.

2.1. System Model

The GNSS multi-interference sources localization scenario in this study is illustrated in Figure 1, which includes both directional and omnidirectional interference sources. The proposed method enables effective GNSS multi-interference sources localization by processing the GNSS data collected from numerous smartphones within the monitored area.

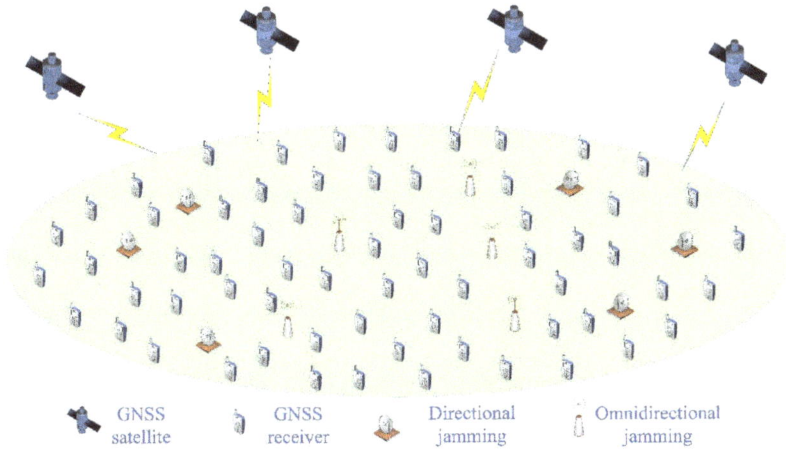

Figure 1. GNSS multi-interference source localization scenario.

In Figure 1, the number of smartphones is M; the C/N_0 returned by the k-th GNSS receiver and its own location information form a three-dimensional vector dataset $\boldsymbol{X}_k = \left[x_k, y_k, (C/N_0)_{eq,k,dB}\right]$, where x_k and y_k are the two-dimensional coordinates of the receiver, and $(C/N_0)_{eq,k,dB}$ is the C/N_0 measurement of the smartphone. This paper utilized the data $\boldsymbol{X}_k, 1 \leq k \leq M$ returned by smartphones in the scenario to confirm the location of interference sources, denoted as $\boldsymbol{C}_i = [x_i, y_i], 1 \leq i \leq N$, with the number of interference sources being N. The aim of this paper is to determine the locations $\boldsymbol{C}_i = [x_i, y_i] (1 \leq i \leq N)$ of multiple interference sources when they coexist, using the data $\boldsymbol{X}_k (1 \leq k \leq M)$ returned by receivers in the scenario. The problem can be described as follows:

$$(\boldsymbol{C}_i, i = 1, \ldots, N) = T(\boldsymbol{X}_k, k = 1, \ldots, M) \tag{1}$$

where T represents the multi-interference sources localization method in this paper. For the scenario described in this paper, the system model was further developed to account for the presence of directional interference sources. Reference [25] provides a method for calculating the C/N_0 of navigation receivers under the simultaneous influence of multiple GNSS interference sources. The formula for calculating the C/N_0 of the k-th receiver in this scenario is as follows.

$$(C/N_0)_{eq,k} = \frac{(C/N_0)\int_{-\infty}^{\infty}\left|H_{R_k}(f)\right|^2 S_f(f)df}{\int_{-\infty}^{\infty}\left|H_{R_k}(f)\right|^2 S_f(f)df + \sum_{i=1}^{N}\frac{J_{i,k}}{N_0}\int_{-\infty}^{\infty}\left|H_{R_k}(f)\right|^2 S_{J_i}(f)S_f(f)df} \tag{2}$$

where $(C/N_0)_{eq,k}$ represents the equivalent C/N_0 of GNSS the k-th receiver under the influence of multiple interferences; C is the satellite signal power; C/N_0 is the C/N_0 without interference; $J_{i,k}$ is the interference signal power of the i-th interference source at the k-th receiver; $H_{R_k}(f)$ is the transfer function of the k-th receiver's filter; $S_f(f)$ is the power spectral density of the satellite signal, normalized to the unit area over infinite bandwidth; and $S_{J_i}(f)$ is the signal power spectral density of interference source i, normalized to unit area over infinite bandwidth. We define the spread-spectrum processing anti-interference quality factor $Q_{i,k}$ related to the interference signal and satellite navigation signal type as:

$$Q_{i,k} = \frac{\int_{-\infty}^{\infty}\left|H_{R_k}(f)\right|^2 S_f(f)df}{R_C\int_{-\infty}^{\infty}\left|H_{R_k}(f)\right|^2 S_{J_i}(f)S_f(f)df} \tag{3}$$

After substituting Equation (3) into Equation (2), Equation (2) can be simplified as follows:

$$(C/N_0)_{eq,k} = \frac{1}{\frac{1}{C/N_0} + \sum_{i=1}^{N} \frac{J_{i,k}/C}{Q_{i,k}R_c}} \tag{4}$$

where the spread spectrum processing anti-jamming quality factor $Q_{i,k}$ is related to the types of interference signals and satellite navigation signals, and R_c represents the pseudocode rate of the satellite guidance signal. Typically, GNSS receivers estimate the C/N_0 in units of $dB \cdot Hz$, and the calculation formula is as follows:

$$(C/N_0)_{eq,k,dB} \triangleq 10lg(C/N_0)_{eq,k} \tag{5}$$

According to the classic narrowband radio propagation path loss model provided in reference [26], the interference signal power $J_{i,k}$ from interference source i at the k-th receiver can be related to the transmission power $J_{i,T}$ of interference source i, as shown in the following equation.

$$(J_{i,k})_{dB} = (J_{i,T})_{dB} - L_0 - 10\alpha lgd - v \tag{6}$$

where $(J_{i,T})_{dB} = 10lg(J_{i,T})$, $(J_{i,k})_{dB} = 10lg(J_{i,k})$, L_0 is the propagation loss at 1 m from the transmitter, d is the distance between transmitter and receiver, α is the path loss index, and v is a Gaussian random variable representing noise. Considering both omnidirectional and directional interference sources, Formula (6) can be expressed as:

$$(J_{i,k})_{dB} = (P_{i,T})_{dB} + (G_{i,k})_{dB} - L_0 - 10\alpha lgd - v \tag{7}$$

where $(P_{i,T})_{dB}$ represents the power of the interference source amplifier, and $(G_{i,k})_{dB}$ denotes the antenna gain of the interference source, which differs between omnidirectional and directional interference. Using the aforementioned model, a simulation scenario was constructed where the positions of the GNSS jammers and GNSS receivers were randomly assigned within a specified area. Using Formula (6), the signal power of each interference source at the location of each GNSS receiver can be calculated in the presence of directional interference sources. Additionally, Formula (4) can be used to calculate the C/N_0 of each GNSS receiver under the influence of multiple interference sources when directional interferences are present.

2.2. C/N_0 Distribution Map Construction Method

Grayscale images are a common type of image in computer vision applications. In a grayscale image, the brightness of each pixel (i,j) is represented by a grayscale value $Gray(i,j)$, which is typically expressed as an 8-bit integer ranging from 0 to 255. In this case, 0 represents black, 255 represents white, and intermediate values correspond to varying shades of gray [27]. Since grayscale images contain only brightness information, they simplify image representation complexity while retaining the essential features of the image. Grayscale images are often used in operations such as edge detection, image segmentation, and feature extraction, enhancing the efficiency and reliability of image processing and analysis [28].

The GNSS receiver information reported by smartphones in Figure 1 includes receiver location data and C/N_0 information, denoted as $\mathbf{X}_k = \left[x_k, y_k, (C/N_0)_{eq,k,dB}\right]$, which exhibits a high degree of similarity to the grayscale information of images. In this study, we utilize the information $\mathbf{X}_k$ relayed by the GNSS receivers embedded in smartphones to construct the C/N_0 distribution map. In this context, the terms "smartphone" and "GNSS receiver" refer to the same device. Therefore, a method is proposed to construct a C/N_0

distribution map by projecting the C/N_0 data collected from numerous GNSS receivers in the monitoring area onto a grayscale image. After generating the C/N_0 distribution map, target detection methods from image processing can be used to detect GNSS interference sources as targets, thereby achieving the effective localization of multiple GNSS interference sources in the scenario of this paper.

Algorithm 1 presents the pseudocode for constructing the C/N_0 distribution map, where the navigation data reported by receivers in the monitored area is used to define the map dimensions and pixel grayscale values. Steps 1 to 3 calculate the receiver density and interference source localization accuracy within the monitored area to obtain the width and height of the C/N_0 distribution map. In this algorithm, the C/N_0 value under interference-free conditions is set to $(C/N_0)_{\max}$, and the receiver's C/N_0 decreases as the level of interference increases [29]. To ensure that the generated C/N_0 distribution map effectively reflects the degree of interference experienced by the receivers, a C/N_0 threshold is set to $(C/N_0)_{CT}$. Grayscale values for C/N_0 values below the threshold $(C/N_0)_{CT}$ are set to 0 and values under interference-free conditions are set to 255, so that the receiver's C/N_0 value $[(C/N_0)_{CT}, (C/N_0)_{\max}]$ can be linearly mapped to grayscale values from 0 to 255 using the following formula:

$$Gray_I = \left\lfloor \frac{255 \cdot [(C/N_0)_I - (C/N_0)_{CT}]}{[(C/N_0)_{\max} - (C/N_0)_{CT}]} \right\rfloor \tag{8}$$

where $(C/N_0)_I$ represents the C/N_0 measurement at a specific receiving point, and $Gray_I$ is the corresponding grayscale value at that point.

Steps 5 to 13 use the C/N_0 measurements reported by the receivers in the monitored area to fill each pixel of the C/N_0 distribution map, in which Step 6 mainly involves searching for data points $\{X_{I,D1}, X_{I,D2}, \ldots, X_{I,DM}\}$ whose distance from the pixel is less than Dp, and DM is the number of points that satisfy the condition. Step 7 calculates the weighted average value $(C/N_0)_{IM}$ of the C/N_0 data for the DM data points, where the weight is the inverse of the distance to that pixel. If $(C/N_0)_{IM} > (C/N_0)_{CT}$, then the gray value corresponding to the value $(C/N_0)_{IM}$ needs to be calculated using Formula (8) and written to the corresponding pixel. If $(C/N_0)_{IM} \leq (C/N_0)_{CT}$, then the gray value 255 is written to that pixel. Steps 14 to 16 apply dilation and median filtering to the generated grayscale image to suppress noise, thereby generating the C/N_0 distribution map.

2.3. Dataset Generation and Problem Description

2.3.1. Dataset Generation

According to the signal model outlined in Section 2 and the construction method of the C/N_0 distribution map described in Section 3.1, a dataset was generated. This dataset comprised 20,000 C/N_0 distribution maps, each corresponding to a single interference localization experiment. For each interference localization experiment, the following parameters were configured:

(1) The monitoring area was set to 60×60 km.

(2) The number of receivers was randomly set between 400 to 600 per square kilometer, with their positions following a random distribution.

(3) The maximum number of interference sources was set to 16. In each trial, the total number of interference sources (denoted as *N_all*) was randomly selected as an integer between 1 and 16. Subsequently, an integer *N_Od* between 1 and *N_all* was randomly chosen to represent the number of omnidirectional interference sources. The number of directional interference sources for that trial was then calculated as *N_all* − *N_Od*. For instance, in one trial, a random integer between 1 and 16 was selected, resulting

in 16 jamming sources. Among these, 7 were randomly designated as omnidirectional jammers, leaving 9 as directional jammers.

(4) The antenna type of each jamming source was randomly assigned using typical models provided by MATLAB R2024a's Antenna Toolbox [30]. Figure 2 illustrates the radiation patterns of different antenna models at the navigation frequency. Among them, the typical omnidirectional antenna is the dipole antenna, while typical directional antennas include Yagi-Uda antennas, horn antennas, Vivaldi antennas, helical antennas, and patch microstrip antennas. Simultaneously, the power of each jammer and the jammer direction were randomly set sequentially. The jammer power was an arbitrary integer between −10 dBm and 10 dBm. The jammer direction was randomly set to an arbitrary angle within the range of 1° to 360°, with an adjustment accuracy of 1°. The jamming transmission was narrowband jamming, corresponding to a Q value of 1.

Algorithm 1: Generation method of the CNR distribution map

INPUT: The number of receivers M
INPUT: Monitoring area width length L and width W
INPUT: The receiver returned data set $\{X_1, X_2, \ldots, X_M\}$
INPUT: The C/N_0 threshold $(C/N_0)_{CT}$
INPUT: The C/N_0 value under interference-free $(C/N_0)_{\max}$
INPUT: Set the distance threshold Dp for selecting C/N_0 data when assigning grayscale values to pixels
OUTPUT: The C/N_0 distribution map ICN
1: Calculate receiver density as $Rd = M/L \times W$
2: Set the localization accuracy as $Lp = 1/\lceil Rd/40 \rceil \times 2$
3: Calculate the width of the C/N_0 distribution map as $W_{ICN} = L/Lp$, the height of the C/N_0 distribution map as $H_{ICN} = W/Lp$, and the number of pixels as $N = W_{ICN} \times H_{ICN}$
4: Initialize the gray values of all pixels in the C/N_0 distribution map ICN to 0 corresponding to the C/N_0 under interference-free conditions
5: While $l \leq N$ do
6: Obtain the receiver returned data set $\{X_{l,D1}, X_{l,D2}, \ldots, X_{l,DM}\}$ that are within a distance of Dp from the pixel
7: Calculate the weighted average $(C/N_0)_{lM}$ of the C/N_0 data for the DM data points, where the weights are the inverse of the distance to the pixel
8: If $(C/N_0)_{lM} > (C/N_0)_{CT}$ then
9: Calculate the grayscale value $Gray_l$ corresponding to the value $(C/N_0)_{lM}$ using Formula (8) and write it to the corresponding pixel
10: Else
11: Write the grayscale value 255 to the corresponding pixel
12: End if
13: End while
14: Perform a dilation operation on ICN
15: Perform a median filtering operation on ICN
16: Return the C/N_0 distribution map ICN

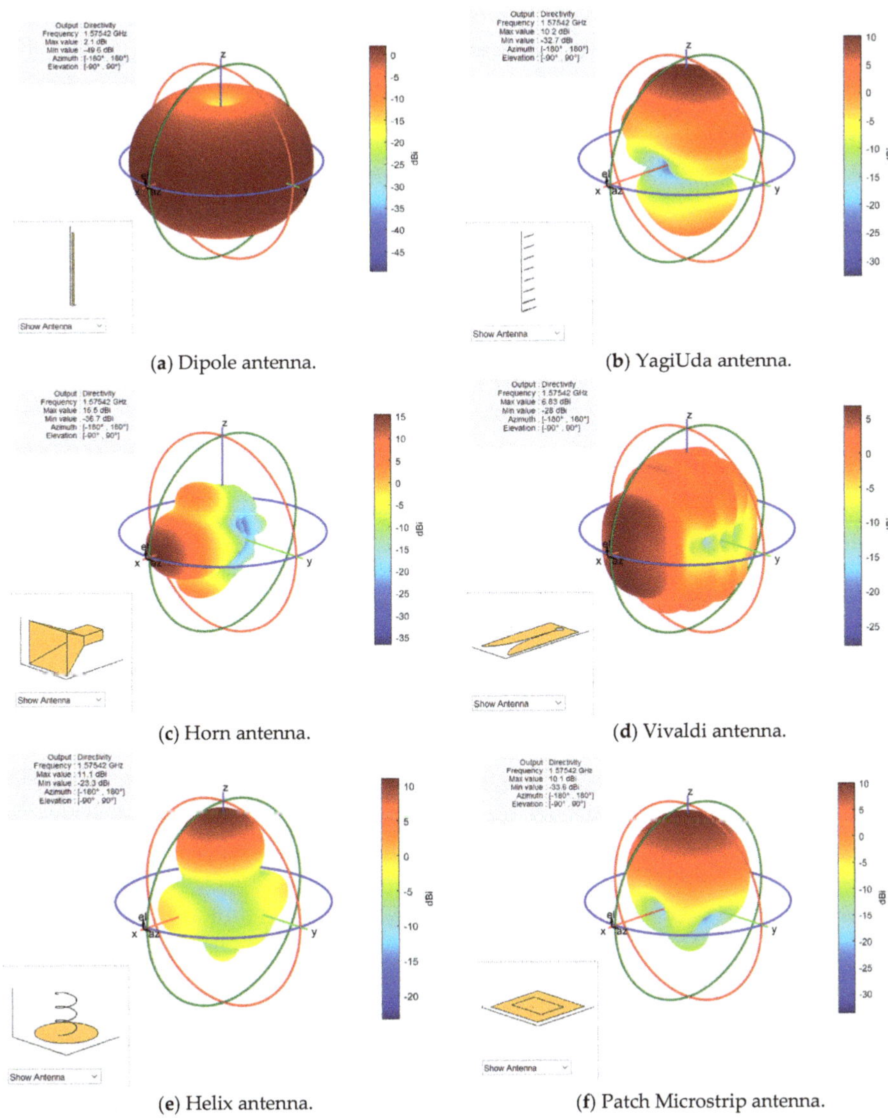

(**a**) Dipole antenna. (**b**) YagiUda antenna. (**c**) Horn antenna. (**d**) Vivaldi antenna. (**e**) Helix antenna. (**f**) Patch Microstrip antenna.

Figure 2. Antenna radiation patterns for different antenna models.

(5) Based on the system model in Section 2, the navigation signal C/N_0 for each receiver in multiple interference source scenarios can be calculated for each experiment. On this basis, the method in Section 3.1 can be used to generate a sample of the C/N_0 distribution map for this experiment.

The dataset generated through the aforementioned steps was utilized for training and testing the proposed target detection method. Representative examples of C/N_0 distribution maps from the dataset are illustrated in Figure 3. As shown, interference effects cause signal C/N_0 degradation, forming interference source shadow targets. Consequently, the interference source localization problem can be transformed into a target detection task within the C/N_0 distribution maps. In these maps, omnidirectional and directional interference sources exhibit distinct image characteristics under different parameter configurations.

For the scenario addressed in this study, the objectives were to identify target types and precisely localize interference sources within the C/N_0 distribution maps. Deep learning-based target detection methods offer significant advantages in terms of both accuracy and computational efficiency. Therefore, this work employed a deep learning-based detection framework to achieve robust interference source detection and localization.

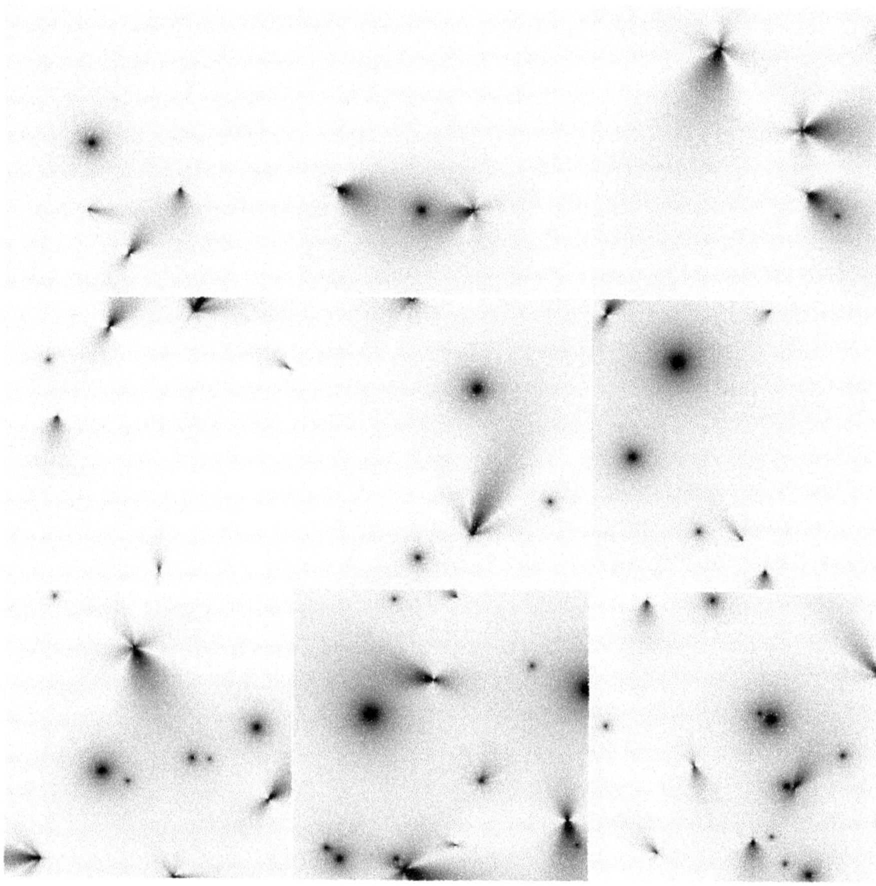

Figure 3. Example of C/N_0 distribution maps from the dataset.

2.3.2. Problem Description

This method in Section 3.1 constructs a C/N_0 distribution map by utilizing the similarity between GNSS receiver return information and image grayscale information. Once the C/N_0 distribution map is obtained, object detection methods from image processing can be employed to detect and locate GNSS interference sources as the targets of interest. Traditional object detection algorithms typically achieve target detection by extracting image features and combining them with classification methods such as Support Vector Machine (SVM). However, these methods are computationally intensive and suffer from poor generalization performance. With the rapid development of deep learning, deep learning-based object detection methods have been widely adopted, offering advantages in accuracy, efficiency, and automation compared to traditional methods. In scenarios such as the coexistence of multiple directional interferences, increased diversity and density of GNSS interference, and the presence of multiple low-power interference sources, further solutions are needed to address the following issues in target localization using the C/N_0 distribution map:

(1) The estimation of the main lobe direction of the directional interference source antenna.

Figure 4 presents illustrative examples of targets in the C/N_0 distribution map under different pointing angles of directional interference using a Horn antenna. As shown, when the interference source uses a directional antenna, its radiation pattern is non-uniform and non-omnidirectional, with the main lobe exhibiting significantly higher gain than other directions. As a result, a shadow region with reduced C/N_0 appears in the direction of the antenna's main lobe. To accurately localize directional interference sources, it is necessary to identify the main lobe direction of the directional antenna. Therefore, the target detection algorithm used must be capable of recognizing rotated targets and outputting their rotation angles.

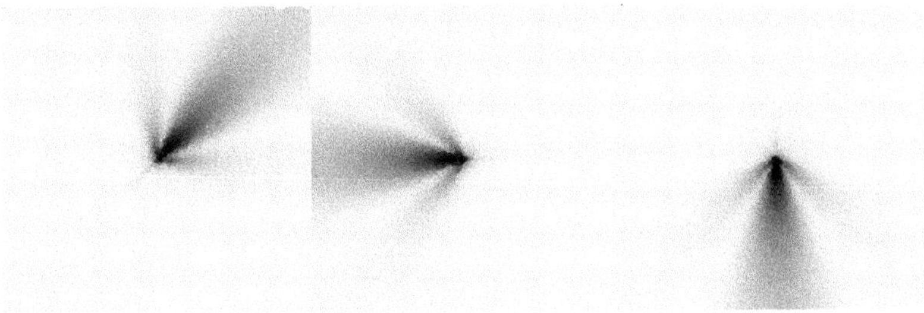

Figure 4. Examples of targets under different antenna pointing angles of Horn antenna interference.

(2) The problem of multi-type antenna target feature recognition and classification.

Figure 5 presents a schematic diagram of different types of interference source antennas at an equivalent radiated power of 0 dBm in the C/N_0 distribution map. As illustrated, interference sources with different antenna types exhibit distinct target characteristics. Therefore, to enable the effective localization of multiple GNSS interference sources under various scenarios, it is essential to first learn the features of each target type and then classify them accordingly.

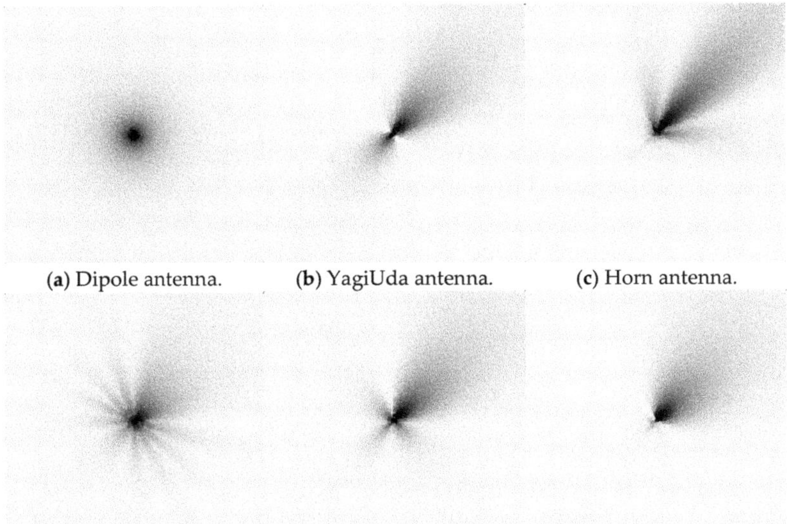

(**a**) Dipole antenna. (**b**) YagiUda antenna. (**c**) Horn antenna.

(**d**) Vivaldi antenna. (**e**) Helix antenna. (**f**) PatchMicrostrip antenna.

Figure 5. Examples of targets from interference sources with different antenna types.

(3) The problem of detecting and localizing weak GNSS interference source targets in the distribution map.

Figure 6 illustrates targets of interference sources with different antenna types at an equivalent radiated power of −15 dBm in the C/N_0 distribution map. As shown, weak GNSS interference sources typically affect only a limited number of navigation receivers, resulting in small-sized targets within the distribution map. To improve the accuracy of detection and localization for such weak interference sources, it is essential to enhance the deep learning model's ability to learn and distinguish the features of small targets associated with various types of weak GNSS interference.

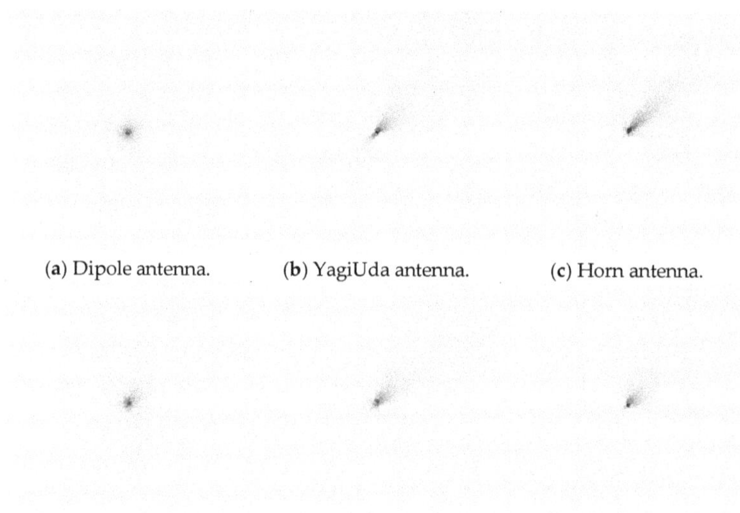

(**a**) Dipole antenna. (**b**) YagiUda antenna. (**c**) Horn antenna.

(**d**) Vivaldi antenna. (**e**) Helix antenna. (**f**) PatchMicrostrip antenna.

Figure 6. Examples of interference source targets with different antenna types at an equivalent radiated power of −15 dBm.

2.4. OSF-RCNN

Deep learning-based object detection methods can be divided into two categories: two-stage approaches based on region proposals and one-stage approaches based on regression. The two-stage method first generates candidate bounding boxes in the processed image, and then classifies these proposals and performs precise positional regression to obtain detection results. Representative algorithms of this approach include the Region-Based Convolutional Neural Network (R-CNN) series [31]. In contrast, one-stage methods treat all possible locations in the image as potential candidate bounding boxes, directly performing object classification and bounding box regression to obtain detection results, with the You Only Look Once (YOLO) series being representative examples [32]. YOLO is a single-stage object detection algorithm that formulates detection as a regression problem. It processes the entire image in a single forward pass and directly predicts bounding boxes along with class probabilities, enabling real-time performance. Since the one-stage method exhibits lower localization accuracy and poorer performance in multi-object detection [33], and this study requires the precise localization of multiple GNSS interference sources, we propose improvements to the Faster R-CNN algorithm to adapt it to the application scenario of this research.

To address the problems outlined in Section 2.3.2, this paper proposes an OSF-RCNN framework based on the Faster R-CNN algorithm for processing C/N_0 distribution maps to achieve the effective localization of multiple GNSS interference sources. Built upon the Faster R-CNN framework, the proposed OSF-RCNN can effectively learn and recognize the features of different types of interference sources. To account for the rotational characteristics of directional antennas, an Oriented RPN was introduced to regress the orientation angles of directional antennas [34]. Additionally, the SE mechanism [35] was integrated into key positions of the network to enhance attention to weak interference signals. Furthermore, FPN was employed to fuse multi-type and multi-scale feature information, thereby improving the detection accuracy of small targets and making the method more suitable for weak GNSS interference detection and localization. The experimental results demonstrate that the proposed method achieves accurate and reliable localization of multiple GNSS interference sources, effectively addressing challenging scenarios such as the presence of multiple directional interferences, increasing diversity and quantity of GNSS interference, and the detection of multiple weak interference signals.

2.4.1. Algorithmic Framework

The overall framework of the proposed OSF-RCNN is illustrated in Figure 7. It first employs a backbone network to extract multi-scale feature maps from the input image. Specifically, the backbone network integrates an SE-enhanced ResNet50 (SE-ResNet50) to extract image features, followed by an FPN that processes and fuses the feature maps output from different stages of the SE-ResNet50. As a result, multi-scale feature maps with rich semantic information and high resolution are generated [36]. Subsequently, an Oriented RPN processes the multi-scale feature maps to generate candidate bounding boxes that encode both direction and rotation information. After obtaining the proposals, both the candidate box information and feature maps are fed into a Rotated RoIAlign layer to achieve precise alignment between the Regions of Interest (RoIs) and the feature maps. Finally, the feature maps corresponding to the candidate boxes are passed through fully connected (FC) layers to perform the classification of the target boxes and regression of their positions and boundaries, thereby achieving accurate target localization.

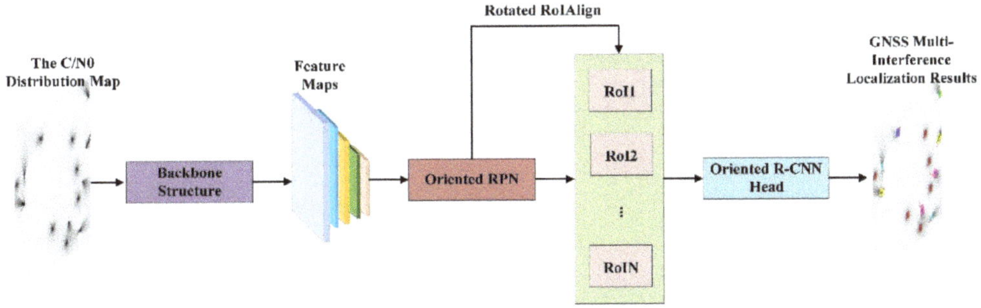

Figure 7. The overall framework of the proposed OSF-RCNN.

2.4.2. Backbone Structure

The backbone network adopts a design that combines SE-ResNet50 with FPN [37], and the overall architecture of the network is shown in Figure 8. ResNet-50 is a deep convolutional neural network architecture renowned for its unique residual learning and efficient training capabilities. It has demonstrated outstanding performance in image classification tasks on benchmark datasets like ImageNet, achieving high accuracy. Due to its residual connection design, ResNet-50 is easier to train and optimize compared to other networks of similar depth [38]. The SE-ResNet50 in the backbone network consists of five stages: Conv1, Conv2_x, Conv3_x, Conv4_x, and Conv5_x. The configuration of each stage is consistent with that described in reference [35]. The outputs of each node in the processing pipeline of the backbone network, which combines SE-ResNet50 and FPN, are shown in Figure 8. This network integrates the feature learning capability of ResNet50 and the multi-scale feature representation ability of FPN, significantly improving target detection performance. This design approach not only enhances the network's adaptability to different interference intensities but also strengthens its ability to extract weak interference features.

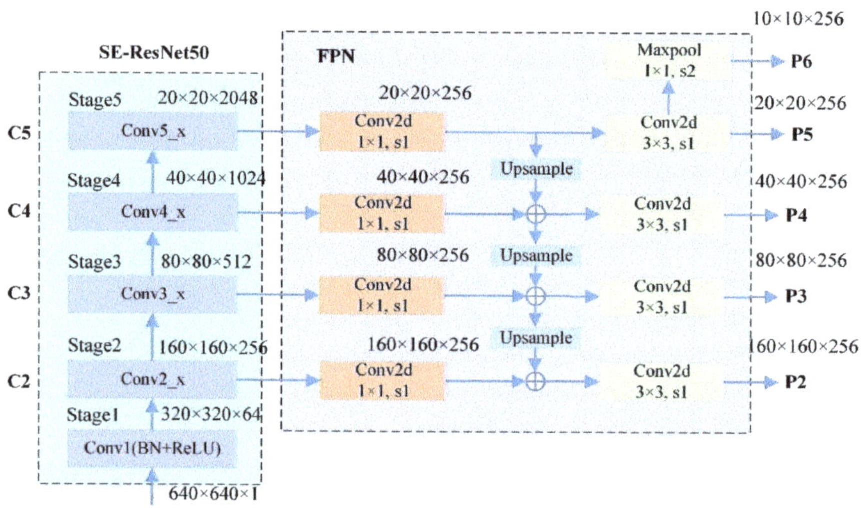

Figure 8. Backbone network architecture.

2.4.3. SE Mechanism

In Section 2.4.2, ResNet-50 consists of two fundamental block structures, the Identity Block and the Conv Block, collectively referred to as the Residual Block. SE-ResNet50 introduces two FC layers into each block of ResNet-50 to recalibrate feature responses, enhancing attention to critical features while suppressing less relevant ones, thereby improving the network's performance more efficiently. This structure is known as the SE mechanism, and its specific architecture is illustrated in Figure 9.

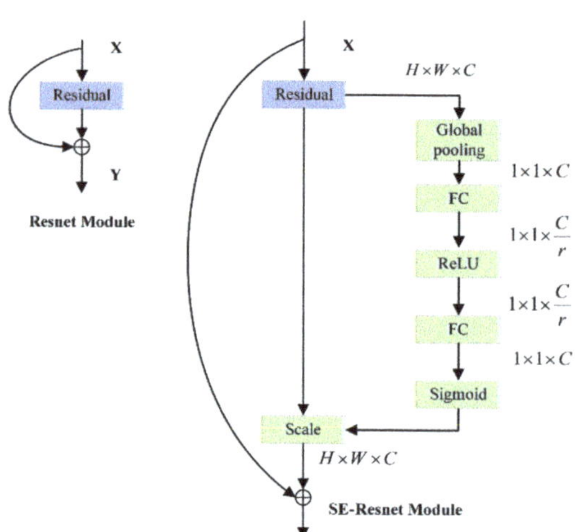

Figure 9. SE mechanism architecture.

The goal of the SE mechanism is to introduce an adaptive mechanism that enables the model to dynamically learn and adjust the importance of each feature channel, thereby enhancing its focus on critical features. The Squeeze operation achieves this by using Global Average Pooling to capture global information for each feature channel. The Excitation operation employs a multilayer perceptron (MLP) to learn excitation weights for the feature channels. This MLP typically consists of two FC layers: the first layer reduces the dimensionality, while the second FC layer restores the reduced dimension back to the original size. The learned excitation weights can be regarded as an attention mechanism that assigns importance to the feature responses of each channel in the feature map. Figure 9 illustrates the processing flow of the SE mechanism on the feature map and the corresponding changes in feature data dimensions. The output of the final Sigmoid function ranges between 0 and 1, representing the importance level of the corresponding feature channel.

2.4.4. Oriented RPN

The Oriented RPN enhances the original RPN by introducing a center-offset representation for rotated bounding boxes. It extends the original four-variable representation to six variables, enabling the inclusion of rotational attributes in the candidate boxes, as illustrated in Figure 10.

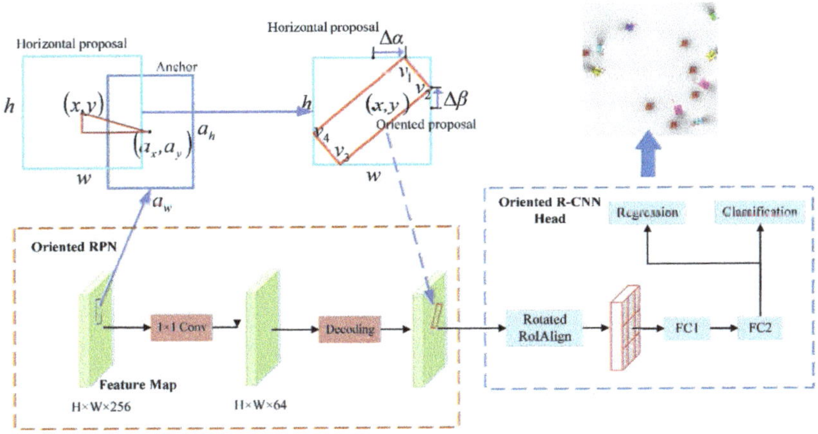

Figure 10. Processing flow of the Oriented RPN and Rotated RoIAlign.

The Oriented RPN improves the regression from anchor boxes to proposal boxes from (x, y, w, h) to $(x, y, w, h, \Delta\alpha, \Delta\beta)$ compared to RPN. The illustration of each variable is shown in Figure 10, where (x, y) represents the center coordinates of the proposal box, w is the width of the horizontal proposal box, h is the height of the horizontal proposal box, and $\Delta\alpha$ and $\Delta\beta$ represent the horizontal and vertical distances from the center of the horizontal proposal to the two vertices of the rotated proposal, respectively. By converting angle regression into distance regression, this approach avoids the non-differentiability of polygonal Intersection over Union (IoU) calculations and the discontinuity issues associated with angular representations.

Based on the above representation, the offset vector between the proposal boxes generated by the Oriented RPN and the anchor boxes is denoted as $(\delta_x, \delta_y, \delta_w, \delta_h, \delta_\alpha, \delta_\beta)$.

From this, the representation vector $(x, y, w, h, \Delta\alpha, \Delta\beta)$ of the rotated proposal can be calculated using the following formulas:

$$\begin{cases} x = a_w \cdot \delta_x + a_x \\ y = a_h \cdot \delta_y + a_y \\ w = a_w \cdot e^{\delta_w} \\ h = a_h \cdot e^{\delta_h} \\ \Delta\alpha = \delta_\alpha \cdot w \\ \Delta\beta = \delta_\beta \cdot h \end{cases} \quad (9)$$

It is important to note that the resulting shape may appear as a parallelogram. To rectify this, the shorter diagonal is extended to equal the length of the longer diagonal. The Oriented RoI generated by the Oriented RPN is processed using the Rotated RoIAlign method. Since the output of Oriented RPN consists of proposals with rotation angles, these proposals need to be rotated into horizontal ones before being aligned. This is the specific operation flow of the Rotated RoIAlign method.

3. Results and Discussion

This section presents a simulation-based analysis of the proposed method. Experiments 1, 2, and 3 evaluate its performance in directional interference scenarios, multi-type interference scenarios, and weak interference scenarios for the detection and localization of multiple GNSS interference sources, respectively. The results demonstrate that the proposed method can achieve the effective localization of GNSS interference sources under these challenging conditions.

3.1. Experiment 1: Simulation of Detection and Localization Performance for Directional Interference Sources

This section primarily evaluates the performance of the proposed method under directional interference scenarios with varying numbers of interference sources. The comparative methods include the approach from Reference [25] and the Faster R-CNN algorithm. Specifically, both Faster R-CNN and the proposed method were evaluated using two representative network architectures: ResNet-50 and VGG-16. ResNet-50 and VGG-16 are currently mainstream backbone networks for image recognition. Both achieve a good balance among accuracy, efficiency, and generalization ability. VGG-16 constructs a deep architecture by continuously stacking small convolutional kernels, featuring a simple and regular structure. ResNet-50 introduces a residual learning mechanism and employs skip connections to effectively alleviate the degradation problem in deep networks [39]. The experimental data were acquired following the methodology described in Section 3.1, with the scenario configuration detailed below:

(1) The GNSS interference monitoring area was set to 60×60 km.

(2) Receiver density varied randomly between 400 and 600 units per square kilometer, with spatial positions assigned randomly.

(3) Two types of interference sources were implemented: standard omnidirectional dipole antennas and directional horn antennas. The transmission power of each source was set to 0 dBm. The direction of each interference source was randomly assigned within the range from 1° to 360°, with a resolution of 1°. All sources emitted narrowband interference signals with a quality factor of Q = 1.

(4) Localization performance was evaluated across varying interference quantities (2–16 sources). For each source count, 100 Monte Carlo trials were conducted with randomized combinations of omnidirectional and directional sources.

(5) The detection accuracy rate acc_{ce} was calculated as $acc_{ce} = T_{ce}/100$, where T_{ce} represents successful detections across 100 trials. Meanwhile, the localization error was also statistically measured when the interference was correctly detected. The localization error was quantified as the mean root mean square error (RMSE) computed across multiple interference localization trials, measured in kilometers. The statistic $RMSE_M$ was calculated using formula $RMSE_M = \sum_{k=1}^{T_{ce}} \sum_{i=1}^{N_{jammmer}} RMSE_i$, where N_{jammer} represents the number of interferences.

In each Monte Carlo trial, the locations of the interference sources, as well as the number and locations of the receivers, were randomly generated. An example of such a trial is illustrated in Figure 11. Figure 11a shows the spatial distribution of the interference sources and receivers, while Figure 11b presents the corresponding C/N_0 distribution map generated based on the configuration in Figure 11a.

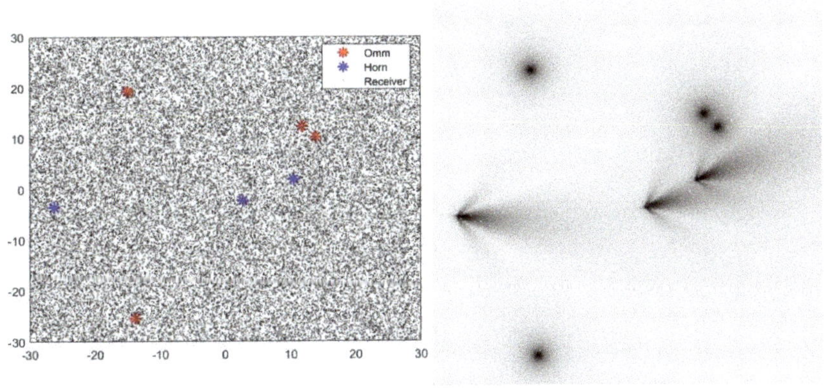

(**a**) (**b**)

Figure 11. An example of a Monte Carlo trial. (**a**) The spatial distribution of the interference sources and receivers. (**b**) The corresponding C/N_0 distribution map.

The experimental results are shown in Figure 12, where Figure 12a presents the statistical results for acc_{ce}, while Figure 12b presents the statistical results for $RMSE_M$. In the figures, CCC represents the method from Reference [25], while our proposed method is denoted as OSF-RCNN. The curves labeled Faster-RCNN-VGG16 and Faster-RCNN-ResNet50 correspond to the Faster R-CNN algorithm implemented with VGG-16 and ResNet-50 backbones, respectively. Similarly, OSF-RCNN-VGG16 and OSF-RCNN-ResNet50 represent the simulation results of our method using VGG-16 and ResNet-50 architectures, respectively.

The simulation results demonstrate that the CCC method exhibits significant degradation in interference detection accuracy as the number of directional interference sources increases, while ResNet-50 outperforms VGG-16 in both interference identification and quantity estimation. When using the ResNet-50 network, both the OSF-RCNN and Faster R-CNN methods can perform interference detection and estimate the number of interference sources, with OSF-RCNN achieving better performance. When different networks are used, the detection and recognition performance of Faster-RCNN-ResNet50 is significantly better than that of OSF-RCNN-VGG16, indicating that the choice of network has a considerable impact on the effectiveness of interference detection and recognition. OSF-RCNN-ResNet50 achieved the best interference detection and localization performance in this experimental scenario. In this experiment, when the number of interference sources reaches 18, the proposed method achieves an interference detection rate acc_{ce} of 77% and a positioning accuracy $RMSE_M$ of 0.5 km.

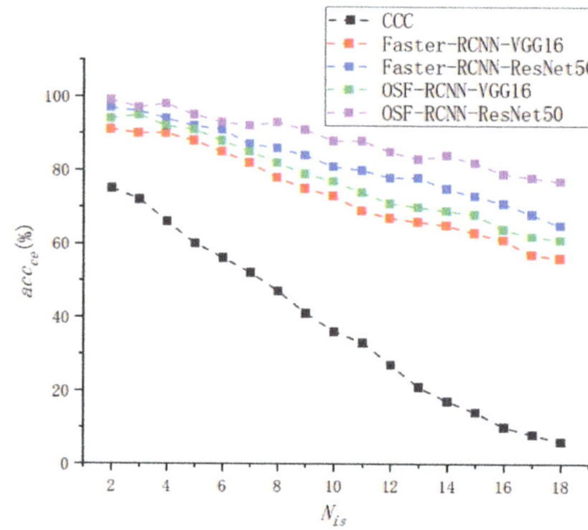

(**a**)

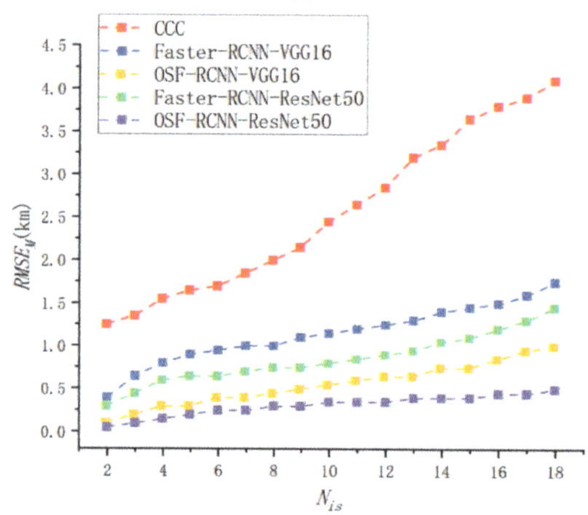

(**b**)

Figure 12. Simulation results for directional interference detection and localization. (**a**) Statistical results of acc_{ce}. (**b**) Statistical results of $RMSE_M$.

Figure 13 presents the experimental comparison between the OSF-RCNN-ResNet50 and OSF-RCNN-VGG16 methods. The OSF-RCNN-VGG16 method misidentifies the directional interference highlighted by the red circle as omnidirectional interference. Compared to the VGG-16 network, the ResNet-50 network has advantages such as a deeper architecture and better gradient propagation, making it more suitable for detecting and localizing directional interference in scenarios involving multiple interference sources.

Figure 14 presents the experimental comparison between the OSF-RCNN-ResNet50 and Faster-RCNN-ResNet50 methods. Figure 14c compares the localized results of the horn directional interference (highlighted in the top-left corners of Figure 14a,b with the ground-truth interference source position). It can be observed that the Faster-RCNN-ResNet50 method fails to accurately estimate the main lobe direction of the directional antenna, resulting in the deviation in the interference source position estimation.

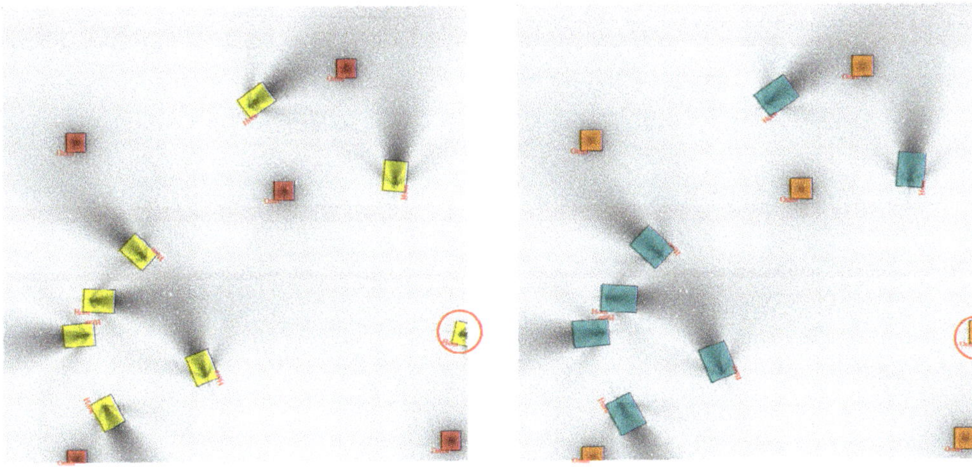

(**a**) (**b**)

Figure 13. Comparison of experimental results between OSF-RCNN-ResNet50 and OSF-RCNN-VGG16 methods. (**a**) Localization performance of OSF-RCNN-ResNet50. (**b**) Localization performance of OSF-RCNN-VGG16.

(**a**) (**b**) (**c**)

Figure 14. Comparison of experimental results between OSF-RCNN-ResNet50 and Faster-RCNN-ResNet50 methods. (**a**) Localization performance of OSF-RCNN-ResNet50. (**b**) Localization performance of Faster-RCNN-ResNet50. (**c**) Localization performance comparison.

As shown in Figure 15, with an increasing number of interference sources, phenomena such as target proximity to the edge of the C/N_0 distribution map and overlap between closely spaced targets lead to aliasing effects. In Figure 5a, where the number of interference sources is six, a localization error of up to 350 m occurs due to a target being near the edge of the C/N_0 map. In another case shown in Figure 15b, with 13 interference sources, localization errors of 538 m and 570 m are caused by overlapping targets. As the number of targets continues to increase, these issues become more pronounced, resulting in further degradation in the localization accuracy.

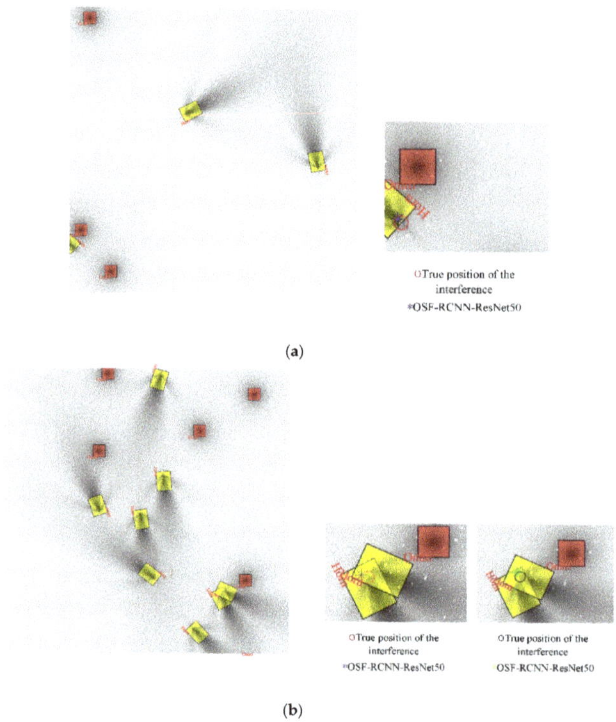

Figure 15. Example of increased localization error. (**a**) The interference target is located at the edge of the image. (**b**) Aliasing occurs due to the close spacing of interference sources.

In conclusion, the CCC method is not applicable for interference source localization in scenarios with directional interference. The OSF-RCNN-ResNet50 method, with ResNet-50 as its backbone network, demonstrates superior performance in interference detection and identification, as well as interference source localization.

3.2. Experiment 2: Simulation of Detection and Localization Performance for Multi-Type Interference Sources

This section primarily evaluates the effectiveness of the proposed method in detecting and localizing multiple interference sources under multi-type interference scenarios using the optimized ResNet-50 network. The experimental data were acquired following the methodology described in Section 3.1, with the scenario configuration detailed below:

(1) The GNSS interference monitoring area was set to 60 × 60 km.

(2) Receiver density varied randomly between 400 and 600 units per square kilometer, with spatial positions assigned randomly.

(3) The number of interference types as set between two and six, with the typical omnidirectional antenna being a dipole antenna. Typical directional antennas include Yagi-Uda, horn, Vivaldi, helix, and patch microstrip antennas. The transmission power of each source was set to 0 dBm. The direction of each interference source was randomly assigned within the range from 1° to 360°, with a resolution of 1°. All sources emitted narrowband interference signals with a quality factor of Q = 1.

(4) The localization performance of the proposed method was tested under each interference type configuration with varying numbers of interference sources. The number of interference sources was set sequentially from 2 to 16, with 100 Monte Carlo trials

conducted for each case. In each trial, the number of omnidirectional and directional interference sources was assigned randomly.

(5) The detection and recognition accuracy of interference sources, acc_{ce}, was calculated, along with the localization error, $RMSE_M$, when the interference was correctly detected and recognized. The statistical methods are consistent with those described in Section 3.1.

The experimental results are shown in Figure 16, where Figure 16a presents the statistical results for acc_{ce}, while Figure 16b presents the statistical results for $RMSE_M$. The curves labeled NoT_2 to NoT_6 correspond to the cases where the number of interference types is set from two to six, respectively. The results demonstrate that the proposed method effectively adapts to complex scenarios with multiple interference types. However, as the number of interference types increases, both detection and localization performance exhibit slight degradation. In this experiment, under the condition of interference type 6 and 18 interference sources, the proposed method achieves an interference detection rate acc_{ce} of 70% and a positioning accuracy $RMSE_M$ of 1.05 km.

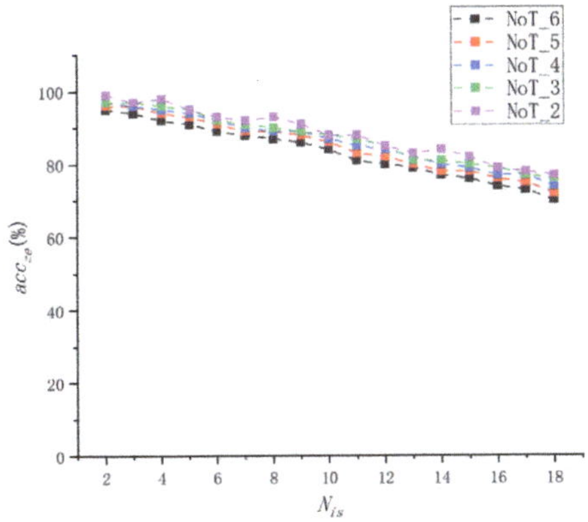

(a)

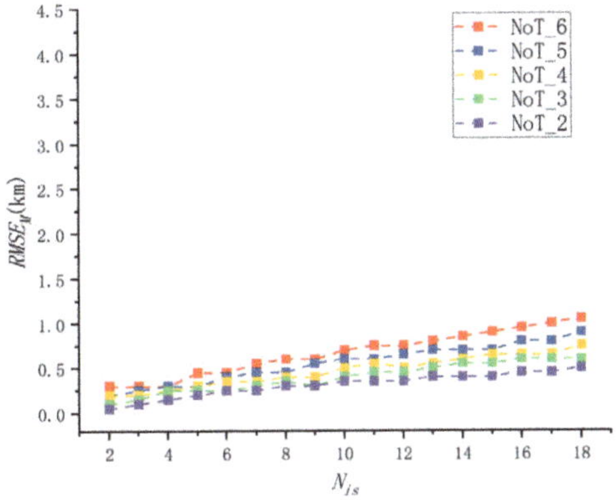

(b)

Figure 16. Simulation results for multi-type interference detection and localization. (**a**) Statistical results of acc_{ce}. (**b**) Statistical results of $RMSE_M$.

Figure 17 illustrates examples of detection and localization performance for multi-type interference. Figure 17a–c show the interference detection and localization results in complex scenarios with interference types set to four, five, and six, respectively. It can be seen that the method proposed in this paper effectively achieves accurate detection and high-precision localization for multiple types of interference. In scenarios where the interference type is set to six and the number of interference sources reaches up to 18, this method can still achieve a detection recognition rate of over 70% and a localization accuracy within 1 km.

Figure 17. Examples of multi-type interference detection and localization performance. (**a**) Detection and localization performance with four interference types. (**b**) Detection and localization performance with five interference types. (**c**) Detection and localization performance with six interference types.

3.3. Experiment 3: Simulation of Detection and Localization Performance Under Varying Interference Transmission Powers

This section evaluates the performance of the proposed method in detecting and localizing multiple interference sources using the optimized ResNet-50 network under varying transmission power levels. The experimental data were acquired following the methodology described in Section 3.1, with the scenario configuration detailed below:

(1) The GNSS interference monitoring area was set to 60 × 60 km.

(2) Receiver density varied randomly between 400 and 600 units per square kilometer, with spatial positions assigned randomly.

(3) The number of interference types was set to three and six, with the typical omnidirectional antenna being a dipole antenna. Typical directional antennas include Yagi-Uda, horn, Vivaldi, helix, and patch microstrip antennas. The direction of each interference source was randomly assigned within the range from 1° to 360°, with a resolution of 1°. All sources emitted narrowband interference signals with a quality factor of Q = 1.

(4) For each interference type setting, the interference source power was set to 0 dBm, −5 dBm, and −15 dBm, respectively.

(5) For each combination of interference type and power, the localization performance of the method was tested with varying numbers of interference sources. The number of interference sources was set sequentially from 2 to 16, with 100 Monte Carlo trials conducted for each case. In each trial, the numbers of omnidirectional and directional interference sources were assigned randomly.

(6) The detection and recognition accuracy of interference sources, acc_{ce}, was calculated, along with the localization error, $RMSE_M$, when the interference was correctly detected and recognized. The statistical methods are consistent with those described in Section 3.1.

The experimental results are shown in Figure 18, where Figure 18a presents the statistical results for acc_{ce}, while Figure 18b presents the statistical results for $RMSE_M$. The result curves under different interference source power levels and interference source type settings were plotted. For instance, 0 dBm_NoT_3 represents the curve where the number of interference source types is three and the interference source power is 0 dBm. The results demonstrate that the proposed method effectively adapts to complex scenarios with weak interference sources. However, as the number of interference types increases and the interference power decreases, the detection and localization performance of the proposed method exhibits a slight decline. As the interference transmission power decreases, changes in transmission power have a greater impact on the localization performance of the proposed method than the increase in the number of interference types. In this experiment, under the condition of an interference power of −15 dBm, interference type six, and 18 interference sources, the proposed method achieves an interference detection rate acc_{ce} of 60% and a positioning accuracy $RMSE_M$ of 1.8 km.

Figure 19 illustrates the interference detection and localization performance under different interference transmission powers when the interference type is set to six. Figure 19a–c show the detection and localization performance of the proposed method when the interference transmission power is 0 dBm, −5 dBm, and −15 dBm, respectively. The method can effectively detect and accurately locate multiple types of weak interference. In a challenging scenario where the interference type is set to six, the interference transmission power is −15 dBm, and the number of interference sources is as high as 18, the proposed method can still achieve a detection recognition rate of over 60% and an interference localization accuracy within 1.8 km.

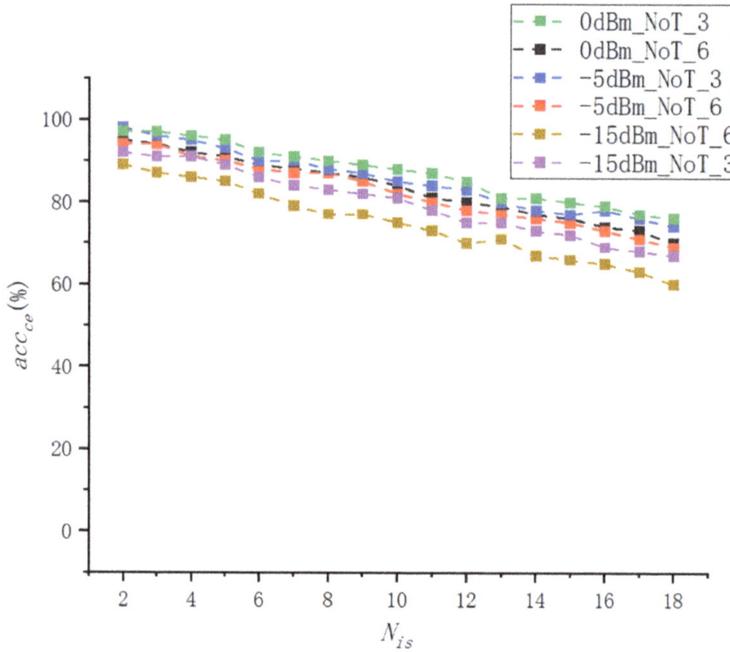

(**a**)

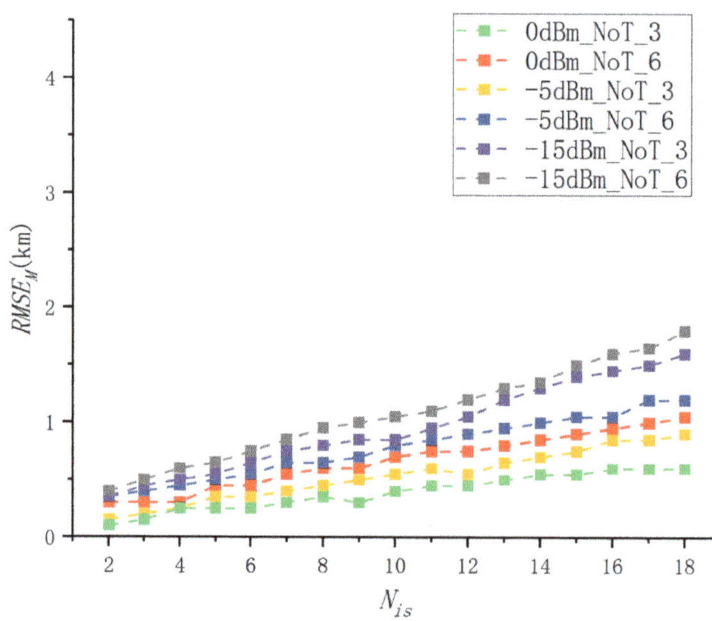

(**b**)

Figure 18. Simulation results under varying interference transmission powers. (**a**) Statistical results of acc_{ce}. (**b**) Statistical results of $RMSE_M$.

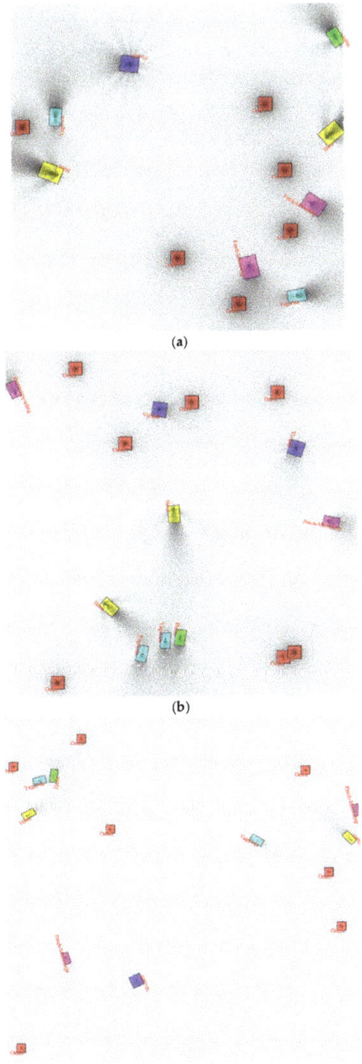

(a) (b) (c)

Figure 19. Interference detection and localization performance under different interference transmission powers when the interference type is set to 6: (**a**) 0 dBm interference transmission power; (**b**) −5 dBm interference transmission power; and (**c**) −15 dBm interference transmission power.

4. Conclusions

This study proposes a multi-interference source localization method based on target detection in GNSS receiver C/N_0 distribution maps, specifically addressing challenging scenarios, including the coexistence of multiple directional interferences, increased diversity and density of GNSS interference, and the presence of multiple low-power interference sources. Building upon established system and signal models suitable for GNSS interference source localization, we innovatively exploit the similarity between GNSS receiver navigation information and image grayscale information to construct C/N_0 distribution maps, thereby transforming the multi-interference source localization problem into a computer vision-based object detection task. To address the specific challenges in target detection and localization for our application scenario, we developed the OSF-RCNN framework by enhancing the Faster R-CNN architecture. The proposed method integrates an Oriented RPN to regress the orientation angles of directional antennas, effectively addressing their rotational characteristics. The FPN is used to fuse multi-scale feature information, enhancing the method's adaptability to different types of interference sources and varying transmission powers. The SE mechanism is introduced at key positions within the network to enhance focus on weak interference targets, improving the detection accuracy of small targets. The paper presents the corresponding system and signal models, constructs training and testing datasets, and conducts simulation tests to evaluate the detection and localization performance of the proposed method. The results demonstrate that the method effectively detects and localizes GNSS interference sources across a range of challenging scenarios.

Author Contributions: Conceptpualization, Q.C. and R.L.; Methodology, Q.C. and R.L.; Software, Q.Y. and X.H.; Validation, R.L. and Q.Y.; Formal analysis, R.L. and X.H.; Investigation, Q.Y., Y.X. and Y.Z.; Resources, Q.C. and Y.L.; Data curation, Y.X. and Y.Z.; Writing—original draft, R.L. and Q.C.; Writing—review and editing, Q.Y. and Q.C.; Project administration, Q.C. and Y.L. All authors have read and agreed to the published version of the manuscript.

Funding: This research received no external funding.

Data Availability Statement: Data are contained within the article.

Conflicts of Interest: The authors declare no conflicts of interest.

Abbreviations

The following abbreviations are used in this manuscript:

GNSS	Global Navigation Satellite Systems
C/N_0	Carrier-to-noise ratio
OSF-RCNN	Oriented Squeeze-and-Excitation-based Faster Region-based Convolutional Neural Network
Faster R-CNN	Faster Region-based Convolutional Neural Network
RPN	Region Proposal Network
SE	Squeeze-and-Excitation
FPN	Feature Pyramid Network
GPS	Global Positioning System
GLONASS	Global Navigation Satellite System
DOA	Direction of Arrival
TDOA	Time Difference of Arrival
RSS	Received Signal Strength
COTS	Commercial off-the-shelf
MIIT	Ministry of Industry and Information Technology
J911	Jamming 911
ION	Institute of Navigation
SVM	Support Vector Machine
RCNN	Region-based Convolutional Neural Network
YOLO	You Only Look Once
ResNet	Residual Network
RoIs	Regions of Interest
FC	Fully connected
RoIAlign	Rotated Region of Interest Align
SE-ResNet50	SE-enhanced ResNet50
MLP	Multilayer perceptron
IoU	Intersection over Union
VGG	Visual Geometry Group
RMSE	Root Mean Square Error

References

1. Kaplan, E.D.; Hegarty, C. (Eds.) *Understanding GPS: Principles and Applications*, 2nd ed.; Artech House Publishers: Norwood, MA, USA, 2005.
2. Borre, K.; Akos, D.; Bertelsen, N.; Rinder, P.; Jensen, S. *A Software–Defined GPS and Galileo Receive: Single–Frequency Approach*; Birkhäuser: Boston, MA, USA, 2007.
3. Dardari, D.; Closas, P.; Djurić, P.M. Indoor tracking: Theory, methods, and technologies. *IEEE Trans. Veh. Technol.* **2015**, *64*, 1263–1278. [CrossRef]
4. Wu, R.; Wang, W.; Lu, D.; Wang, L.; Jia, Q. *Adaptive interference mitigation in GNSS*; Springer: Singapore, 2018.
5. Wang, H.; Yao, Z.; Yang, J.; Fan, Z. A Novel Beamforming Algorithm for GNSS Receivers with Dual-Polarized Sensitive Arrays in the Joint Space–Time-Polarization Domain. *Sensors* **2018**, *18*, 4506. [CrossRef] [PubMed]
6. Balaei, A.T.; Motella, B.; Dempster, A.G. GPS interference detected in Sydney-Australia. In Proceedings of the IGNSS Conference, Sydney, Australia, 4–6 December 2007; pp. 1–12.
7. Kuusniemi, H.; Airos, E.; Bhuiyan, M.Z.H.; Kröger, T. GNSS jammers: How vulnerable are consumer grade satellite navigation receivers. *Eur. J. Navig.* **2012**, *10*, 14–21.
8. Dempster, A.G.; Cetin, E. Interference Localization for Satellite Navigation Systems. *Proc. IEEE* **2016**, *104*, 1318–1326. [CrossRef]
9. Bhamidipati, S.; Gao, G.X. GPS Multireceiver Joint Direct Time Estimation and Spoofer Localization. *IEEE Trans. Aerosp. Electron. Syst.* **2018**, *55*, 1907–1919. [CrossRef]
10. Gao, G.X.; Sgammini, M.; Lu, M.; Kubo, N. Protecting GNSS Receivers from Jamming and Interference. *Proc. IEEE* **2016**, *104*, 1327–1338. [CrossRef]
11. Rounds, S. Innovation-Jamming Protection of GPS Receivers-Part II: Antenna Enhancements-The second article in a two-part series about jamming protection for GPS receivers looks at enhancements to antenna. *GPS World* **2004**, *15*, 38–45.
12. Daneshmand, S.; Jafarnia-Jahromi, A.; Broumandan, A.; Lachapelle, G. A GNSS structural interference mitigation technique using antenna array processing. In Proceedings of the 2014 IEEE 8th Sensor Array and Multichannel Signal Processing Workshop, A Coruna, Spain, 22–25 June 2014.
13. Zhang, J.; Cui, X.; Xu, H.; Lu, M. A Two-Stage Interference Suppression Scheme Based on Antenna Array for GNSS Jamming and Spoofing. *Sensors* **2019**, *19*, 3870. [CrossRef] [PubMed]
14. Amin, M.G.; Wang, X.; Zhang, Y.D.; Ahmad, F.; Aboutanios, E. Sparse Arrays and Sampling for Interference Mitigation and DOA Estimation in GNSS. *Proc. IEEE* **2016**, *104*, 1302–1317. [CrossRef]
15. Fresno, J.M.; Robles, G.; Martínez-Tarifa, J.M.; Stewart, B.G. Survey on the Performance of Source Localization Algorithms. *Sensors* **2017**, *17*, 2666. [CrossRef] [PubMed]
16. Amin, M.G.; Borio, D.; Zhang, Y.D.; Galleani, L. Time-Frequency Analysis for GNSSs: From interference mitigation to system monitoring. *IEEE Signal Process. Mag.* **2017**, *34*, 85–95. [CrossRef]
17. Borio, D.; Gioia, C.; Štern, A.; Dimc, F.; Baldini, G. Jammer Localization: From Crowdsourcing to Synthetic Detection. In Proceedings of the 29th International Technical Meeting of The Satellite Division of the Institute of Navigation, Manassas, VA, USA, 12–16 September 2016; pp. 3107–3116. [CrossRef]
18. Wang, P.; Morton, Y.T. Efficient Weighted Centroid Technique for Crowdsourcing GNSS RFI Localization Using Differential RSS. *IEEE Trans. Aerosp. Electron. Syst.* **2020**, *56*, 2471–2477. [CrossRef]
19. Mills, K.; Ahmad, F.; Amin, M.G.; Himed, B. Fast Iterative Interpolated Beamforming for Interference DOA Estimation in GNSS Receivers Using Fully Augmentable Arrays. In Proceedings of the 2019 IEEE Radar Conference (RadarConf), Toulon, France, 23–27 September 2019; pp. 1–5. [CrossRef]
20. Chen, Q.D.; Tao, H.H.; Liu, R.; Zhen, W.M. Time Difference Estimation Method for TDOA Location of GNSS Weak Interference. *J. China Acad. Electron. Inf. Technol.* **2020**, *15*, 135–140. (In Chinese) [CrossRef]
21. Khandker, S.; Torres-Sospedra, J.; Ristaniemi, T. Analysis of Received Signal Strength Quantization in Fingerprinting Localization. *Sensors* **2020**, *20*, 3203. [CrossRef] [PubMed]
22. Statistical Bulletin of the Communications Industry of the Central People's Government of the People's Republic of China in 2023. Available online: https://www.gov.cn/lianbo/bumen/202401/content_6928019.htm (accessed on 1 June 2025).
23. Strizic, L.; Akos, D.M.; Lo, S. Crowdsourcing GNSS Jammer Detection and Localization. In Proceedings of the 2018 International Technical Meeting of The Institute of Navigation, Reston, VA, USA, 29 January–1 February 2018; pp. 626–641.
24. Scott, L. J911: The case for fast jammer detection and location using crowdsourcing approaches. In Proceedings of the 24th International Technical Meeting of The Satellite Division of the Institute of Navigation (ION GNSS 2011), Portland, OR, USA, 19–23 September 2011; pp. 1931–1940.
25. Liu, R.; Yang, Z.; Chen, Q.; Liao, G.; Zhen, W. GNSS multi-interference source centroid location based on clustering centroid convergence. *IEEE Access* **2021**, *9*, 108452–108465. [CrossRef]
26. Shafi, M.; Takada, J. Advances in Propagation Modeling for Wireless Systems. In *EURASIP Journal on Wireless Communications and Networking*; Hindawi Publishing Corporation: London, UK, 2009; p. 5.

27. Burger, W.; Burge, M.J. *Digital Image Processing: An Algorithmic Introduction*; Springer Nature: London, UK, 2022.
28. Eddins, S.L.; Gonzalez, R.C.; Woods, R.E. *Digital Image Processing Using Matlab*; Princeton Hall Pearson Education Inc.: Hoboken, NJ, USA, 2004; pp. 6–12.
29. Xie, G. *Principles of GPS and Receiver Design*; Publishing House of Electronics Industry: Beijing, China, 2009; pp. 244–246.
30. Zou, Z.; Chen, K.; Shi, Z.; Guo, Y.; Ye, J. Object detection in 20 years: A survey. *Proc. IEEE* **2023**, *111*, 257–276. [CrossRef]
31. Mittal, P.; Singh, R.; Sharma, A. Deep learning-based object detection in low-altitude UAV datasets: A survey. *Image Vis. Comput.* **2020**, *104*, 104046. [CrossRef]
32. Chen, C.; Zheng, Z.; Xu, T.; Guo, S.; Feng, S.; Yao, W.; Lan, Y. YOLO-Based UAV Technology: A Review of the Research and Its Applications. *Drones* **2023**, *7*, 190. [CrossRef]
33. Yan, D.; Li, G.; Li, X.; Zhang, H.; Lei, H.; Lu, K.; Chen, M.; Zhu, F. An improved faster R-CNN method to detect tailings ponds from high-resolution remote sensing images. *Remote Sens.* **2021**, *13*, 2052. [CrossRef]
34. Xie, X.; Cheng, G.; Wang, J.; Yao, X.; Han, J. Oriented R-CNN for object detection. In Proceedings of the IEEE/CVF International Conference on Computer Vision, Montreal, QC, Canada, 10–17 October 2021; pp. 3520–3529.
35. Hu, J.; Shen, L.; Sun, G. Squeeze-and-excitation networks. In Proceedings of the IEEE Conference on Computer Vision and Pattern Recognition, Salt Lake City, UT, USA, 18–23 June 2018; pp. 7132–7141.
36. Lin, T.Y.; Dollár, P.; Girshick, R.; He, K.; Hariharan, B.; Belongie, S. Feature pyramid networks for object detection. In Proceedings of the IEEE Conference on Computer Vision and Pattern Recognition, Honolulu, HI, USA, 21–26 July 2017; pp. 2117–2125.
37. Kirillov, A.; He, K.; Girshick, R.; Dollár, P. A unified architecture for instance and semantic segmentation. In Proceedings of the IEEE Conference on Computer Vision and Pattern Recognition, Honolulu, HI, USA, 21–26 July 2017.
38. Cai, Z.; Vasconcelos, N. Cascade r-cnn: Delving into high quality object detection. In Proceedings of the IEEE Conference on Computer Vision and Pattern Recognition, Salt Lake City, UT, USA, 18–23 June 2018; pp. 6154–6162.
39. Wu, F.; Lin, H. Effect of transfer learning on the performance of VGGNet-16 and ResNet-50 for the classification of organic and residual waste. *Front. Environ. Sci.* **2022**, *10*, 1043843. [CrossRef]

remote sensing

MDPI

Article

Basis Recovery Method for Ionospheric Delay Corrections in PPP-RTK Model with Recommendations for Interpolation Reference Station Number Selection

Siyao Wang [1], Runzhi Zhang [1,2,*], Rui Tu [3], Lihong Fan [1] and Xiaochun Lu [1,2]

1 National Time Service Center, Chinese Academy of Sciences, Xi'an 710600, China; wangsiyao@ntsc.ac.cn (S.W.); fanlihong@ntsc.ac.cn (L.F.); luxc@ntsc.ac.cn (X.L.)
2 University of Chinese Academy of Sciences, Beijing 100049, China
3 College of Geodesy and Geomatics, Shandong University of Science and Technology, Qingdao 266590, China; turui@sdust.edu.cn
* Correspondence: zhangrunzhi@ntsc.ac.cn

Abstract: Precise point positioning–real-time kinematic (PPP-RTK) enables users to achieve rapid centimeter-level absolute positioning accuracy within a few epochs. The interpolation of ionospheric delay corrections at the user end, extracted from reference stations, constitutes a key aspect of the process, which depends not solely on the precision of the interpolation model. This study investigates the recommended number of selected reference stations and proposes a method to mitigate the potential loss of observations due to missing ionospheric corrections. According to the experimental results, the number of reference stations should be determined based on the reference network size. Under normal conditions (terrain is relatively flat and the atmospheric conditions are inactive) where reference stations are approximately evenly distributed in all directions, and using low-order surface interpolation model, for networks with 50 km spacing, four or five reference stations are recommended, while for 100 km networks, six or seven stations are enough to calculate precise corrections. Adding more stations beyond these thresholds provides limited improvement in interpolation accuracy and increases the communication load. In addition, an interpolation basis recovery algorithm is proposed to preserve otherwise excluded satellite observations through intelligent handling of correction data gaps at individual reference stations. Experimental validation demonstrates that the recovered ionospheric delay corrections obtained through the algorithm deviate from the ground-truth interpolated values of no more than ±1 cm, an accuracy level deemed adequate for PPP-RTK applications. Furthermore, approximately 3% of the observations, which would otherwise have been discarded due to the missing corrections from a specific reference station, are retained by the algorithm.

Keywords: global navigation satellite system; precise point positioning; interpolation of ionospheric delays; reference network density; interpolation basis recovery

1. Introduction

Precise point positioning (PPP) [1,2] has become increasingly widespread in various applications, including navigation, geodesy, and time and frequency transfer. In recent years, the PPP with ambiguity resolution (PPP-AR) technique has addressed the impact of uncalibrated phase delays (UPDs), which are difficult to separate in float PPP solutions, on the ambiguities [3]. The methods for ambiguity resolution are flexible and include the integer recovery clock (IRC) method [4], the decoupled clock method [5], and the

UPD method [6]. These methods demonstrate consistency in both theoretical aspects and practical performance [7]. Furthermore, to achieve instantaneous ambiguity resolution (IAR), precise point positioning–real-time kinematic (PPP-RTK) has been proposed, which facilitates rapid ambiguity fixing by correcting atmospheric delays based on a regional reference network [8,9]. Li et al. [10,11] proposed a regional augmented PPP (RAPPP) model based on the UPD method to achieve IAR. Some scholars have also proposed the undifferenced and uncombined PPP-RTK models based on the *S*-system theory [12–14]. The performance of PPP-RTK in achieving instantaneous convergence to a centimeter-level static solution across several epochs has been validated [10–14]. Meanwhile, research and practical applications of low-cost and Internet of Things (IoT)-based location devices in PPP-RTK [15,16] indicate that the technique has broad potential for use in a wide range of fields. Furthermore, satellite-based broadcast is expected to offer promising prospects for future technological applications [17].

The influence of ionospheric delay has become an increasingly critical factor in the field of GNSS positioning research [18,19]. The interpolation of atmospheric delay corrections, especially ionospheric delay corrections, is a critical step in obtaining high-precision delay corrections for users in PPP-RTK applications. Numerous interpolation models have been extensively researched and validated by a variety of scholars. Representative interpolation models are the distance-based linear interpolation method (DIM) [20], the linear combination model (LCM) [21], the low-order surface model (LSM) [22], and others. Wang et al. [23] conducted a comprehensive comparison of the interpolation accuracy of these methods, revealing that their accuracies are nearly identical. However, when addressing the problem of atmospheric delay corrections interpolation in non-specific network environments, the quality of the interpolated corrections is determined not only by the choice of interpolation model but also by several other factors. With the increasing sophistication of reference station networks, selecting the appropriate reference stations has become a critical issue, especially when multiple available reference stations are available in the vicinity of the users. Typically, users select several nearby reference stations for interpolation; however, when redundant reference stations are available in the surrounding area, the proper number of reference stations to use remains uncertain and has yet to be definitively established.

Two critical factors must be considered in determining the number of reference stations to be selected. First, increasing the number of reference stations does not necessarily significantly improve interpolation precision. Due to the inherent accuracy limitations of the interpolation model, once the number of stations exceeds a certain threshold, further increases yield only marginal improvements in the precision of interpolated atmospheric corrections. Additionally, due to the high-frequency data transmission and coordinate updates in PPP-RTK, an unlimited increase in the number of reference stations would impose a significant burden on both communication and data storage. Furthermore, it may introduce inconsistencies in the interpolated corrections, as some reference stations may lack corrections for certain satellites, leading to discrepancies in the basis of the interpolated corrections for different satellites.

Second, the scales of the networks should be considered when the problem of selecting the number of reference stations is discussed. The required number of reference stations varies between smaller networks, where the average station spacing is 30 to 50 km, and larger networks, where the spacing exceeds 100 km. For larger-scale reference station networks, increasing the number of stations used for interpolation can provide additional redundant information, enhancing the accuracy of approximating the atmospheric conditions over a wider region. To develop a proper scheme for selecting the number of reference stations in networks of different scales, this study performed interpolation experiments using varying numbers of reference stations across networks of various scales. The precision of the interpolated corrections was evaluated, and recommendations for the number of reference stations were provided.

Besides the selection of reference stations, the loss of atmospheric corrections from reference stations remains a potential issue in practical PPP-RTK scenarios, which can negatively impact the overall performance of PPP-RTK. An analysis of the atmospheric delay corrections streams recorded during the real-time PPP-RTK service reveals that the absence of certain corrections is a relatively frequent occurrence. The atmospheric delay corrections incorporate hardware delays (or systematic biases) related to the receivers of individual reference stations, which are stable for each receiver [24,25]. The extracted atmospheric delay corrections include a compounded hardware delay term, reflecting the receivers of all reference stations involved in the interpolation process. Generally, atmospheric delay corrections for each satellite are interpolated using the same reference stations, resulting in corrections of each satellite containing the same combined receiver-related hardware delay component. During PPP calculation processing, the receiver clock offset parameter absorbs the uniform hardware delay present in all satellite interpolation corrections, meaning it does not affect the estimation of ambiguity or other parameters. Meanwhile, the uniform receiver-related hardware delay can be calibrated in engineering applications [26].

Conversely, the absence of satellite corrections from reference stations poses a potential risk to this process. These missing corrections can be categorized into two types. The first type involves the complete absence of corrections from a specific reference station, which may result from issues such as communication failures, power outages, or hardware malfunctions. The second type involves missing corrections for specific satellites at individual reference stations, often caused by problems such as signal dropouts, cycle slips, outliers, or ambiguity resolution failures. These instances are typically temporary and intermittent. Once the receiver re-establishes tracking and phase lock with the satellite, normal correction generation for that satellite is usually recovered. However, in the case of missing corrections from a specific reference station for a certain satellite, the satellite can only use the remaining reference stations for interpolation of corrections. In contrast, other satellites that have not experienced such missing corrections can use all reference stations. As a result, the receiver-related basis generated by the interpolated corrections for this satellite will differ significantly from that of the other satellites. The receiver clock offset cannot absorb both receiver-related bases simultaneously. Therefore, in traditional methods, if the corrections from a reference station are missing, the interpolated corrections are treated as outliers, and the corresponding satellite observation is discarded. To avoid wasting observations and ensure the continuous availability of corrections throughout the observation period, this study analyzes the impact of both types of correction loss on data processing and proposes a practical and effective basis recovery method. It should be noted that another major factor affecting the reliability of corrections in practical measurements is the time delay (latency) occurring in the communication link. This aspect is not the focus of this paper and has not been considered or addressed herein. Common methods for handling this issue can be found in Odijk [27] and Khodabandeh [28].

The subsequent sections first analyze the effects of the two types of correction loss and then present a detailed explanation of the proposed method, supported by equations. The experimental section is organized into two parts: validation of the proposed method and exploration of the recommended reference station selection strategy. Finally, a discussion and summary are provided.

2. Materials and Methods

This section first provides a brief overview of the generation of ionospheric delay corrections and the conventional interpolation methods for PPP-RTK users. Following this, a solution for cases of missing corrections is subsequently proposed.

2.1. Methods for the Generation and Interpolation of Ionospheric Delay Corrections

The service center extracts the atmospheric corrections from each reference station, compiles them, and transmits them to PPP-RTK users for interpolation and to correct atmospheric delay errors. The reference stations independently perform continuous dual-frequency ionosphere-free (IF) ambiguity resolution, allowing ionospheric delay corrections to be extracted from the resolved ambiguity solutions [25]. The equation derivation here starts with the completion of parameter estimation and ambiguity resolution at the reference stations. Once a reference station achieves continuous wide-lane [29,30] and narrow-lane ambiguity resolution, the ionospheric delay can be directly extracted from the uncombined raw code and phase observations, which is expressed as:

$$\begin{cases} \Delta I_{i,P}^{s} = \rho^{s} - ct_{\mathrm{IF}}^{s} + ct_{\mathrm{IF}} + T^{s} - P_{i}^{s} \\ \Delta I_{i,\Phi}^{s} = \rho^{s} - ct_{\mathrm{IF}}^{s} + ct_{\mathrm{IF}} + T^{s} + \lambda_{i} N_{i}^{s} + \lambda_{i}\mathrm{UPD}_{i}^{s} - \lambda_{i}\phi_{i}^{s} \end{cases} \quad (1)$$

where P and ϕ represent the raw code and phase observations, respectively; subscripts i and s denote the frequency and satellite; ρ is the receiver-to-satellite geometric distance, which is known for reference stations; c is the speed of light; T^{s} denotes the tropospheric delay; t_{IF} and t_{IF}^{s} are the receiver and satellite IF clock offset based on the International GNSS Service (IGS) satellite clock offset products, respectively; N and λ represent the integer phase ambiguity and its wavelength; $\Delta I_{i,P}^{s}$ and $\Delta I_{i,\Phi}^{s}$ are the extracted code and phase ionospheric delay corrections; and UPD_{i}^{s} represents the UPD corrections in cycles, which contain phase and code hardware delays. The UPD productions have already been estimated by the service center based on the reference network, and the code hardware delays in UPDs will be absorbed in the user observations along with the phase ionospheric delay corrections [25,31,32]. Moreover, after the ambiguities are correctly fixed, the integer-ambiguity-constrained parameters, such as the tropospheric delay, clock offset, etc., can be acquired. Consequently, the values of ρ^{s}, δ, t_{IF}^{s}, t_{IF}, T^{s}, and N_{i}^{s} can be regarded as prior values. As a result, the estimated code and phase ionospheric corrections become available.

Both code and phase ionospheric delay corrections include receiver-related hardware delays. Therefore, the interpolation basis recovery algorithm is applicable to both. Taking the phase ionospheric delay correction for satellite s as an example, assume that the n surrounding reference stations have individually generated the phase ionospheric delay corrections for satellite s. The user can then receive corresponding phase ionospheric delay corrections, i.e., $\left[\Delta I_{\Phi}^{s}[1]\ \Delta I_{\Phi}^{s}[2]\ \ldots\ \Delta I_{\Phi}^{s}[n]\right]$. $\Delta I_{\Phi}^{s}[n]$ represents the ionospheric delay correction extracted from reference station n. Subsequently, the user can correct the ionospheric errors related to spatial and temporal variations by interpolating the ionospheric corrections from the reference stations. For reference station networks across various scales and terrains, the low-order surface model, which incorporates two horizontal parameters, one height parameter, and a constant term, demonstrates the greatest applicability among several commonly used models [23]. The model denoted as H3V1, which will be adopted in this work, can be expressed as follows:

$$\Delta I = a_0 + a_1 \Delta X + a_2 \Delta Y + a_3 \Delta H \quad (2)$$

where ΔI represents the interpolated ionospheric delay corrections at the user station; a_i represents the i-th interpolation coefficient; $(\Delta X\ \Delta Y\ \Delta H)$ represents the difference in the horizontal and vertical coordinates between the user station and the n-th reference station, respectively. In the experimental section of this study, the H3V1 model is used for interpolation of corrections when the number of reference stations is at least four. However, when only three reference stations are available, the simplified H3V0 model is employed, in which the default height parameter is omitted. This model can be viewed as a further simplification of the H3V1 model. Detailed information on the formulation and assumptions of H3V0 can be found in [23]. The H3V0 model is expressed as follows:

$$\Delta I = a_0 + a_1 \Delta X + a_2 \Delta Y \tag{3}$$

Subsequently, the parameter a_i of the H3V0 model can be computed from the n reference stations using the following equation:

$$\begin{bmatrix} 1 & \Delta X_{1u} & \Delta Y_{1u} \\ 1 & \Delta X_{2u} & \Delta Y_{2u} \\ \vdots & \vdots & \vdots \\ 1 & \Delta X_{nu} & \Delta Y_{nu} \end{bmatrix} \begin{bmatrix} a_0 \\ a_1 \\ a_2 \end{bmatrix} = \begin{bmatrix} \Delta I^s_\Phi[1] \\ \Delta I^s_\Phi[2] \\ \vdots \\ \Delta I^s_\Phi[n] \end{bmatrix} \tag{4}$$

where u represents the user station, and $(\Delta X_{nu}\ \Delta Y_{nu})$ represents the coordinate difference between the n-th reference station and the user station u.

2.2. Basis Recovery Method

Due to the fact that the precision of phase corrections is significantly higher than that of range corrections, and they exert a much greater influence on the estimation results, the equation derivation in this section takes phase corrections as an example, while the processing for range and phase corrections is consistent. At reference station n, the phase ionospheric delay correction for satellite s is composed of four components: the actual ionospheric delay, receiver-related hardware delay, satellite-related hardware delay, and noise, which is expressed as [25]:

$$\Delta I^s_\Phi[n] = \overset{\sim s}{I}_\Phi[n] + d^s + d[n] + \varepsilon^s_\Phi[n] \tag{5}$$

where $\Delta I^s_\Phi[n]$ represents the ionospheric delay correction extracted from reference station n; $\overset{\sim s}{I}_\Phi[n]$ is the actual ionospheric delay at reference station n; $d[n]$ and d^s represent receiver-related and satellite-related hardware delay, respectively; and $\varepsilon^s_\Phi[n]$ represents the noise. The interpolation model in Equations (2) and (3) can be expressed as a linear combination of the corrections from each reference station [23], as follows:

$$\Delta I^s_\Phi[u] = \sum_{r=1}^{n} \beta_r \Delta I^s_\Phi[r] \tag{6}$$

where β represents the interpolation coefficients of the model, and subscript u and r represent the user station and reference station. By substituting into Equation (5), the following expression is obtained:

$$\Delta I^s_\Phi[u] = \sum_{r=1}^{n} \beta_r \overset{\sim s}{I}_\Phi[r] + \sum_{r=1}^{n} \beta_r d[r] + d^s + \sum_{r=1}^{n} \beta_r \varepsilon^s_\Phi[r] \tag{7}$$

where the constraint on the model coefficients, $\sum_{r=1}^{n} \beta_r = 1$, is upheld, and superscript n represents the number of reference stations. Due to this constraint, d^s remains unchanged before and after interpolation. Meanwhile, due to inherent errors in the interpolation model, $\sum_{r=1}^{n} \beta_r \overset{\sim s}{I}_{\Phi}[r]$ does not precisely equal $\overset{\sim s}{I}_{\Phi}[u]$. The deviation can be expressed in terms of the interpolation model error $m_n^s[u]$, which is expressed as:

$$\sum_{r=1}^{n} \beta_r \overset{\sim s}{I}_{\Phi}[r] = \overset{\sim s}{I}_{\Phi}[u] + m_n^s[u] \tag{8}$$

where $m_n^s[u]$ represents the interpolation model error associated with both the reference stations and satellites. Upon neglecting the noise term, Equation (7) can be expressed as:

$$\Delta I_{\Phi}^s[u] = \overset{\sim s}{I}_{\Phi}[u] + m_n^s[u] + \sum_{r=1}^{n} \beta_r d[r] + d^s \tag{9}$$

where $m_n^s[u]$ is unavoidable and can only be minimized by increasing the number of reference stations and utilizing more precise interpolation models. Missing atmospheric corrections received by the users can be categorized into two types: the first type occurs when corrections for all satellites from particular reference stations fail to reach the user, while the second type arises when corrections for specific satellites from one or more reference stations are missing.

2.2.1. Missing Corrections for All Satellites from One or More Specific Reference Stations

This situation is similar to the case where the user changes reference stations as their position changes, and it necessitates an analysis of the parameters affected in Equation (9). Assume that in the previous epoch, atmospheric corrections were interpolated using n reference stations. In the current epoch, the user does not receive corrections from p of these stations and can only interpolate using the remaining $n-p$ reference stations. The impact on the ionospheric delay interpolation in both epochs, as outlined in Equation (9), is reflected in the following:

$$\begin{cases} \Delta m = m_n^s[u] - m_{n-p}^s[u] \\ \Delta d = \sum\limits_{r=1}^{n} \beta_r d[r] - \sum\limits_{r=1}^{n-p} \gamma_r d[r] \end{cases} \tag{10}$$

where Δm is the influence of the model error, which is unavoidable, and γ represents the interpolation coefficients of the model with $n-p$ stations. However, when the number of reference stations is sufficiently large, the magnitude of the model error is typically small enough to be neglected. The impact of receiver-related hardware delay Δd, which is independent of the satellite, will be directly absorbed by the receiver clock offset parameter, thus having no effect on the parameter estimation, particularly the ambiguity parameters. Moreover, the impact lies in the fact that the receiver clock offsets between the two epochs introduce different hardware delays, which do not affect the receiver clock offset estimation in the presence of white noise.

In summary, the change in the set of reference stations caused by the complete loss of atmospheric corrections for all satellites from particular reference stations does not impact parameter estimation and does not require separate handling.

2.2.2. Missing Corrections for Specific Satellites from a Particular Reference Station

In cases of missing corrections, when the number of surrounding reference stations is sufficiently large and the distances are close enough, the situation can be treated similarly to a complete loss of corrections, where reference stations with incomplete corrections are excluded. However, if the number of surrounding reference stations is limited, excluding these stations would clearly have a detrimental effect on the quality of the interpolated corrections.

Assuming that, in the current epoch, due to missing corrections, satellite s can only be interpolated using $n-p$ reference stations, while the remaining satellites can still utilize all n reference stations for interpolation. The receiver-related delay in the correction for satellite s is $\sum_{r=1}^{n-p} \gamma_r d[r]$, while for the other satellites, it is $\sum_{r=1}^{n} \beta_r d[r]$. This discrepancy means that the error cannot be absorbed by the receiver clock offset. Consequently, the variation in the receiver hardware delay reference for an individual satellite will inevitably affect the other estimated parameters, particularly the ambiguity parameters. These parameters will be compelled to absorb a portion of the receiver-related hardware delay, preventing their resolution during subsequent ambiguity resolution processes.

In traditional PPP-RTK algorithms, when corrections for specific satellites from certain reference stations are missing, the corrections from anomalous reference stations must be excluded to maintain unbiased estimation of the ambiguity parameters, and uncorrected observations cannot be utilized. As a result, the observations with erroneous corrections can be discarded. This approach inevitably leads to the loss of valuable observational data. To address the problem, we propose a basis recovery method for anomalous reference station corrections. The core idea is to repair the anomalous receiver-related basis, allowing the corrections to be used normally. It is important to note that the method is designed for occasional missing corrections from individual satellites at specific reference stations. When corrections for a satellite are missing from multiple reference stations, it is most likely due to satellite-related issues (such as satellite maneuvers, hardware failures, or low elevation angles). In such cases, the success rate and reliability of ambiguity resolution for that satellite are low, and failures or errors in fixing will significantly degrade the positioning solution quality. Therefore, it is not recommended to use this method. Instead, the satellite observations should be discarded. Therefore, the method is limited to the scenario where only one reference station is missing corrections, i.e., $p=1$. The steps for recovering the basis of the missing corrections are as follows:

(1) Assume that corrections for satellite s are unavailable from the n-th reference station. For the remaining $s-1$ satellites, for which corrections from all reference stations are accessible, the user's corrections $\left[\Delta I_{\Phi}^{1}[u]\ \Delta I_{\Phi}^{2}[u]\ \ldots\ \Delta I_{\Phi}^{s-1}[u]\right]$ are computed by utilizing data from all available reference stations with Equation (1).

(2) Calculate the corrections for the $s-1$ satellites using $n-1$ reference stations, denoted as $\left[\Delta \bar{I}_{\Phi}^{1}[u]\ \Delta \bar{I}_{\Phi}^{2}[u]\ \cdots\ \Delta \bar{I}_{\Phi}^{s-1}[u]\right]$, where the correction for the i-th satellite is written as:

$$\Delta \bar{I}_{\Phi}^{i}[u] = \overset{\sim i}{I}_{\Phi}[u] + m_{n-1}^{i}[u] + \sum_{r=1}^{n-1} \gamma_r d[r] + d^{i} \tag{11}$$

where $m_{n-1}^{i}[u]$ represents the model error for satellite i when interpolating using $n-1$ reference stations; the constraint on the model coefficients, $\sum_{r=1}^{n-1} \gamma_r = 1$, is upheld using $n-1$ reference stations.

(3) Calculate the difference between the two sets of corrections. Based on Equations (9) and (11), the difference between the two types of corrections for satellite i can be denoted as:

$$\nabla \Delta I_{\Phi}^{i}[u] = \Delta I_{\Phi}^{i}[u] - \Delta \bar{I}_{\Phi}^{i}[u] = m_{n}^{i}[u] - m_{n-1}^{i}[u] + \sum_{r=1}^{n} \beta_r d[r] - \sum_{r=1}^{n-1} \gamma_r d[r] \tag{12}$$

where $m_{n}^{i}[u] - m_{n-1}^{i}[u]$ represents the model error term, denoted as $\Delta m_{n,n-1}^{i}$; the constraint on the model coefficients, $\sum_{r=1}^{n} \beta_r = 1$, is upheld using n reference stations; and $\Delta m_{n,n-1}^{i}$ is satellite-dependent, difficult to model and predict, and may exhibit varying signs and magnitudes for different satellites. In the algorithm, it is simplified as a random variable; $\sum_{r=1}^{n} \beta_r d[r] - \sum_{r=1}^{n-1} \gamma_r d[r]$, denoted as $\Delta d_{n,n-1}$, represents the receiver hardware delay at the reference station, which is satellite-independent and a slowly varying error term.

(4) Considering that model errors are random and difficult to estimate accurately, averaging data from multiple satellites can effectively mitigate the impact of these errors. By smoothing the model error using averaging, the difference in receiver hardware delays between the two sets of corrections is calculated. The calculation process can be represented as:

$$\begin{aligned} \frac{\sum_{i=1}^{s-1}\left(\nabla\Delta I_{\Phi}^{i}[u]\right)}{s-1} &= \frac{\sum_{i=1}^{s-1}\left(m_{n}^{i}[u]-m_{n-1}^{i}[u]+\sum_{r=1}^{n}\beta_{r}d[r]-\sum_{r=1}^{n-1}\gamma_{r}d[r]\right)}{s-1} \\ &= \frac{\sum_{i=1}^{s-1}\Delta m_{n,n-1}^{i}}{s-1} + \sum_{r=1}^{n}\beta_{r}d[r] - \sum_{r=1}^{n-1}\gamma_{r}d[r] = \Delta d_{n,n-1} \end{aligned} \tag{13}$$

where $\Delta m_{n,n-1}^{i}$ is considered a random parameter; the term $\frac{\sum_{i=1}^{s-1}\Delta m_{n,n-1}^{i}}{s-1}$ is considered to be quite small after smoothing (can be ignored in the equation expression), which facilitates the calculation of $\Delta d_{n,n-1}$.

(5) For the satellite s with missing corrections, the corrected value is obtained by applying the receiver hardware delay difference calculated in the previous step as the correction. The process is as follows:

$$\Delta I_{\Phi}^{s}[u] = \Delta \bar{I}_{\Phi}^{s}[u] + \Delta d_{n,n-1} = \overset{\sim s}{I}_{\phi}[u] + m_{n-1}^{s}[u] + d^{s} + \sum_{r=1}^{n}\beta_{r}d[r] \tag{14}$$

where for satellite s, the model error $m_{n-1}^{s}[u]$ cannot be corrected. However, when $n-1$ reference stations are sufficient to be used to interpolate the accurate errors, the difference in model errors is small enough to be neglected, i.e., $m_{n-1}^{s}[u] \approx m_{n}^{s}[u]$. By substituting $m_{n-1}^{s}[u]$ in Equation (14) with $m_{n}^{s}[u]$, it becomes consistent with Equation (9). The corrected phase ionospheric correction for satellite s, as derived from Equation (14), can then be considered a normal correction and will no longer affect parameter estimation and ambiguity resolution. The effectiveness and reliability of the basis recovery method will be analysed in the subsequent experimental section. The complete flowchart of the basis recovery method is illustrated in Figure 1.

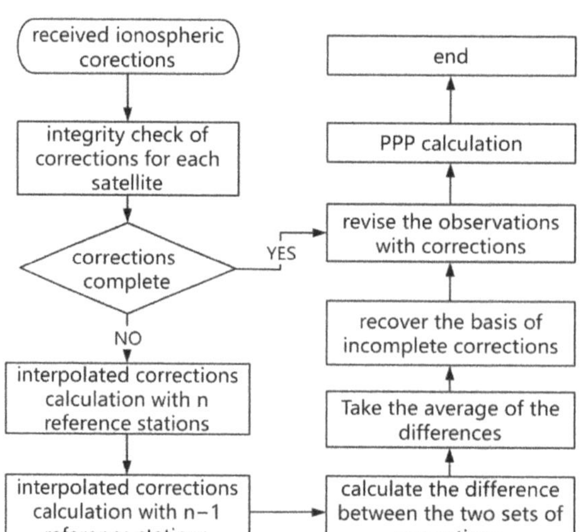

Figure 1. Flowchart of the basis recovery method.

3. Results

The dataset and the experimental results are presented in this subsection, along with the related discussion. The results of the experiment on the basis recovery algorithm and the experiment on the recommended number of reference stations are given in sequence.

3.1. Dataset

The experimental dataset was obtained from a continuously operating reference stations (CORS) network in North Carolina, USA, consisting of 26 stations in total. The regional station network was selected because the terrain in this area is flat, the density of the station network is relatively high, and the stations are distributed relatively uniformly. This enables the convenient selection of stations to form a network, thereby meeting the networking requirements for multiple experiments. Among these, 24 stations were designated as reference stations, while 3 stations were used to simulate user stations. Data were collected for the entire day of 9 April 2024, from 00:00 to 24:00, with a sampling interval of 5 s, and included dual-frequency observations from both the GPS and Galileo systems. The locations of the stations are shown in Figure 2. The ionospheric conditions during the observation period were stable, with no evident solar activity or large-scale geomagnetic changes. The signal-to-noise ratio and the rate of cycle slips in the observations were within normal ranges. It should be noted that the terrain of this area is relatively flat, and the atmospheric conditions on this day were inactive. These environmental conditions serve as the foundation for ensuring the validity and accuracy of conclusions. Different observational environments can potentially lead to deviations in experimental results. The experiment was conducted in two components: one for the validation of the basis recovery algorithm, and the other for determining the recommended number of reference stations.

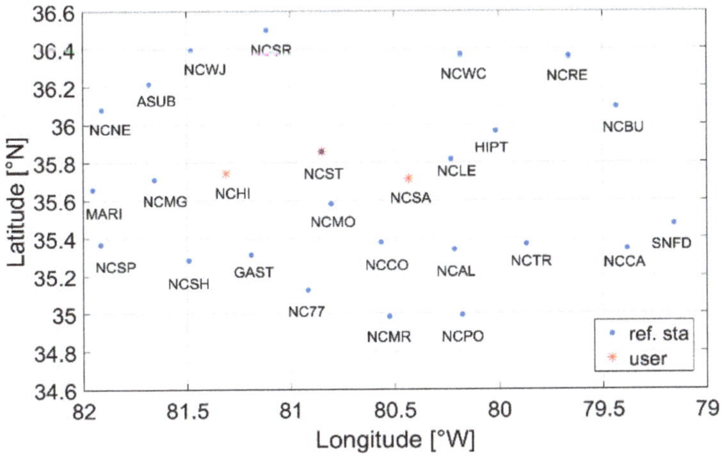

Figure 2. Geographical distribution of the stations (blue dots represent reference stations, red asterisks represent user stations).

3.1.1. Experimental Data for the Basis Recovery Algorithm

The experiment involved four reference stations (NCLE, NCST, NCMO, and NCCO) and a user station (NCSA). The experiment assumes that the user station receives corrections for all satellites from the NCLE, NCST, and NCCO stations, while certain satellites are unable to obtain corrections from the reference station NCMO.

The basis of the corrections from reference station anomalies was recovered using the proposed method. These recovered corrections were then compared with those obtained by interpolation from all four reference stations to assess the feasibility of the proposed method. As the number of reference stations increases, the interpolation model error $m_n^i[u] - m_{n-1}^i[u]$ tends to decrease. To assess the lower bound of the proposed method's reliability, the experiment was conducted using the minimum number of reference stations (only four). In this experiment, communication is assumed to be in a seamless state, meaning that potential unexpected long communication delays are not considered.

To further demonstrate the performance of the proposed method in practical PPP-RTK positioning, a PPP-RTK positioning experiment has been designed in the experiment section to compare the differences before and after applying the proposed method. Firstly, we present the performance of the proposed method on the Position Dilution of Precision (PDOP) throughout the day at the NCSA station. Subsequently, a PPP-RTK experiment is designed. To comprehensively show the ambiguity resolution performance at different startup times, the PPP engine is initiated every ten minutes starting from 0:20 to 23:30, utilizing 30 min of observational data for each run. The epoch interval is set at 5 s, and an attempt is made to fix ambiguities at each epoch. The evaluation indicators include the ambiguity fixing rate and the ratios for epochs where ambiguity resolution is successfully achieved. The ambiguity fixing rate is defined as the ratio of the number of epochs with fixed ambiguities to the total number of solved epochs.

3.1.2. Experimental Data for Assessing the Number of Reference Stations

As discussed earlier, the determination of the proper number of reference stations is inherently linked to the scale of the network. To comprehensively evaluate and provide recommendations for the proper number of reference stations, experiments were conducted using two large-scale networks with station spacings exceeding 100 km and two medium-to-small-scale networks with spacings between 30 and 50 km. The details of the four networks and the simulated rover stations are provided in Tables 1–4, which also include the average distances between the reference and user stations. In addition, the distribution of reference stations can also affect the results of correction interpolation. Therefore, when reference station networks with different numbers of stations are selected, they are distributed as evenly as possible in all directions to achieve the greatest generalizability. Consequently, the applicability of the experiment conclusions will be limited to the premises, representing the most common and general conditions, namely, a flat terrain, an inactive period of the ionosphere, and evenly distributed reference stations.

In the experimental section, the interpolated corrections for two GNSS systems were computed for each of the four networks. The corrections generated by the user station itself were used as the reference values for accuracy evaluation. This study focuses on the interpolation of ionospheric corrections, while the impact of inaccurate interpolated corrections on subsequent positioning results has been previously discussed in Wang et al. [25].

Table 1. A small-scale reference network with NCHI as the user station receives dual-frequency signals from the GPS/Galileo systems (Network A).

Number of Reference Stations	Reference Stations	Mean Reference-User Distance [km]
3	ASUB, NCMO, NCMG	47.7
4	3 + NCSH	49.2
5	4 + GAST	49.1
6	5 + NCST	48.2
7	6 + MARI	49.8

Table 2. A small-scale reference network with NCSA as the user station receives dual-frequency signals from the GPS/Galileo systems (Network B).

Number of Reference Stations	Reference Stations	Mean Reference-User Distance [km]
3	NCLE, NCST, NCAL	36.1
4	3 + NCMO	36.2
5	4 + HIPT	38.5
6	5 + NCCO	38.6
7	6 + NCTR	42.2

Table 3. A large-scale reference network with NCST as the user station receives dual-frequency signals from the GPS/Galileo systems (Network C).

Number of Reference Stations	Reference Stations	Mean Reference-User Distance [km]
3	NCBU ASUB, NCSP	108.6
4	3 + NCCA	117.6
5	4 + NCMR	114.3
6	5 + NCWC	109.1
7	6 + NCSR	104.3
8	7 + NC77	103.4

Table 4. A large-scale reference network with NCSA as the user station receives dual-frequency signals from the GPS/Galileo systems (Network D).

Number of Reference Stations	Reference Stations	Mean Reference-User Distance [km]
3	SNFD, NCNE, NCPO	113.7
4	3 + NCRE	110.3
5	4 + NCWJ	112.3
6	5 + GAST	107.3
7	6 + NCMG	107.8
8	7 + NCSP	117.8

3.2. Experiment on the Basis Recovery Algorithm

This section first explores the feasibility of the basis recovery algorithm and demonstrates its performance. It should be noted that for the situation where all corrections are missing for one or more reference stations, a simple and clear explanation has already been provided in the method section. Therefore, the experimental section will only verify the situation where corrections for individual satellites are missing. It is assumed that the ionospheric delay corrections for all satellites at NCLE, NCST, and NCCO stations are successfully received, while the ionospheric delay corrections for each satellite at the NCMO station may be subject to reception failure. For each satellite at the user station NCSA, interpolation is performed for both scenarios: one where the NCMO station (with four reference stations) can provide the corrections, and the other where it cannot (with three reference stations). Due to differences in the frequencies and tracking methods of GPS and Galileo observations, the receiver-related hardware delays differ between the two systems. Consequently, the performance of the basis recovery algorithm is shown and evaluated separately for each system. The difference between the two sets of interpolated corrections is then calculated, and the results are statistically analyzed. Figures 3 and 4 show the differences in L1 ionospheric corrections for all satellites in the GPS and Galileo systems, respectively.

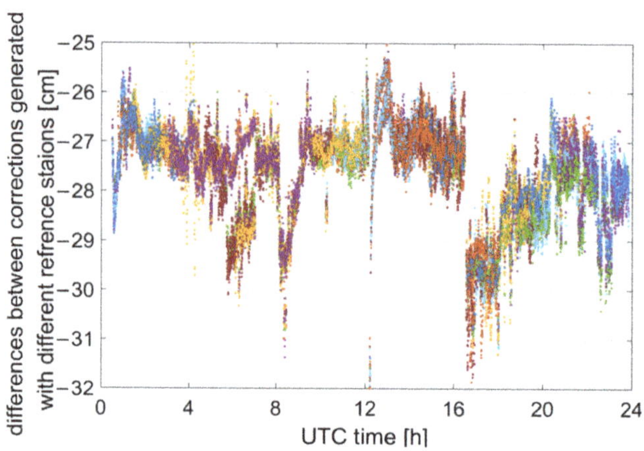

Figure 3. Difference in interpolated L1 ionospheric corrections for all satellites in the GPS system using three and four reference stations.

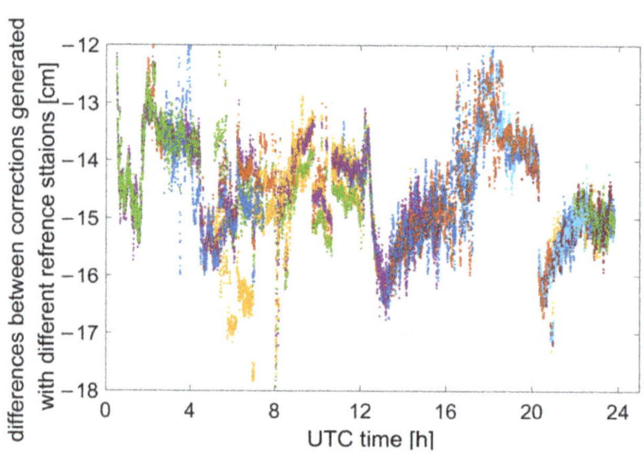

Figure 4. Difference in interpolated L1 ionospheric corrections for all satellites in the Galileo system using three and four reference stations.

Based on Equation (12), the results of the two interpolations encompass both the interpolation model error $\Delta m^i_{n,n-1}$ and the difference in the linearly combined receiver hardware delay $\Delta d_{n,n-1}$. Due to differences in observation frequencies and tracking methods between systems, $\Delta d_{n,n-1}$ exhibits system-dependent biases across different systems. Taking the results for all GPS satellites in Figure 3 as an example, the combined receiver hardware delays $\Delta d_{n,n-1}$ are consistent across all satellites, with the variations primarily governed by the interpolation model error $\Delta m^i_{n,n-1}$. The interpolation model errors $\Delta m^i_{n,n-1}$ for all satellites are mostly confined to within 1 cm. The maximum error is approximately 2 to 3 cm, and its impact on ambiguity resolution is relatively limited given the wavelength of about 20 cm. Therefore, the basis recovery method of deriving $\Delta d_{n,n-1}$ through mean smoothing is supported by these findings. It should be clarified here that the interruptions observed in Figures 3 and 4 were caused by the failure to fix ambiguities at multiple reference stations during those periods. This led to the inability to generate accurate atmospheric corrections. It is speculated that these failures might have been triggered by fluctuations in the quality of UPD products.

After extracting and compensating $\Delta d_{n,n-1}$ using the basis recovery algorithm, Figure 5 shows the difference between the L1 recovered ionospheric delay corrections for all satellites at the NCSA station using the proposed method with three reference stations and the ground-truth interpolated corrections with four reference stations. The ground-truth corrections refer to the corrections computed by fixing the coordinates of the reference stations to their known precise values. Figure 5 illustrates the different results for all GPS and Galileo satellites, with the systematic biases extracted and corrected. The recovered corrections differ from the ground-truth interpolated values only in $\Delta m^{i}_{n,n-1}$, with errors remaining within a 2 cm fluctuation. The basis recovery method, however, preserves the valuable observations that would otherwise be discarded.

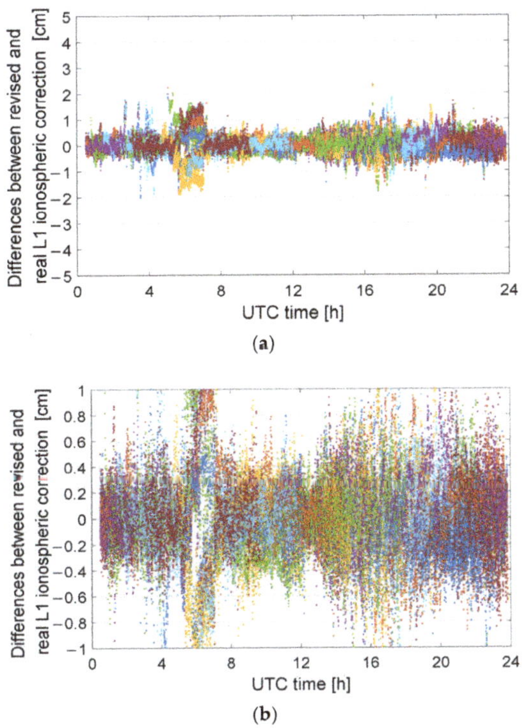

Figure 5. Difference between the L1 recovered ionospheric delay corrections for all satellites at the NCSA station using three reference stations and the ground-truth interpolated corrections with four reference stations. (**a**) The range of the vertical axis is ±5 cm. (**b**) The range of the vertical axis is ±1 cm.

Furthermore, Figure 6 presents the probability distribution of the difference between the recovered corrections for all satellites at the NCSA station using the proposed method with three reference stations and the ground-truth interpolated corrections with four reference stations. Compared to the conventional discard strategy, the basis recovery algorithm generates a novel ionospheric delay correction, thereby improving the utilization of observations in PPP-RTK. The resulting ionospheric delay corrections deviate from the ground-truth interpolated values by no more than ±1 cm. This level of accuracy is adequate for PPP-RTK users [11].

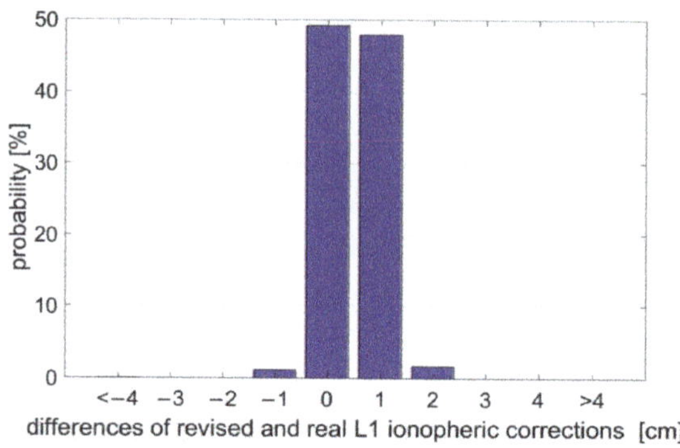

Figure 6. Probability distribution of the difference between the recovered corrections and the true corrections.

Subsequently, a detailed account of the basis recovery method's performance in the PPP-RTK positioning experiment will be provided. To begin with, the primary advantage of the proposed method lies in its ability to reinstate the usability of observations without corrections. When contrasted with the scenario where the method is not employed, there is an increase in the number of available observations. As a result, the satellite geometric distribution configuration becomes more favorable. Table 5 presents the proportion of epoch numbers with PDOP variations, as well as the PDOP improvement observed before and after the implementation of the proposed method.

Table 5. Changes in DOP values before and after using the basis recovery method.

GNSS System	Proportion of Epochs with Varying DOP [%]	Average Difference in PDOP (Without Basis Recovery)	Average Difference in PDOP (With Basis Recovery)	Proportion of Reduction in PDOP After Basis Recovery [%]
Single GPS	5.16	1.84	1.58	14.1
GPS/Galileo	8.79	1.16	1.08	6.9

As indicated in Table 5, when utilizing the single GPS system and the GPS/Galileo system separately, the proportions of epochs with varying DOP are 5.16 and 8.79, respectively. Moreover, the PDOP declined to 1.58 and 1.08, corresponding to reduction rates of 14.1% and 6.9%, respectively. The aforementioned data offers a clear illustration of the proportions of epochs impacted by the proposed method, as well as the gains achieved in DOP values. Actually, when users are located in areas where GNSS signals are obstructed, i.e., in environments where the number of observable satellites significantly decreases, the additional observation information provided by the proposed method will offer greater benefits. In order to illustrate the minimum level of gains that this method can offer, the experiment was conducted without intentionally creating an obstructed setting. Rather, observation data from an open environment were employed. Table 5 illustrates the distinct gain patterns for single-system and multi-system configurations. It reveals that as the quantity of observable satellites increases, the additional observational data has a diminishing impact on the DOP gain. Consequently, if users find themselves in signal-constrained regions with a reduced number of observable satellites, the DOP gains they achieve are expected to exceed the results presented in Table 5.

The PPP-RTK experiment was conducted following the experiment design described in the dataset section. Table 6 displays the ambiguity fixing rate along with the average ratio for epochs where ambiguity has been successfully fixed (with a ratio greater than 2).

Table 6. Ambiguity fixing rate along with the average ratio of PPP-RTK experiment.

GNSS System	Ambiguity Fixing Rate (Without Basis Recovery) [%]	Ambiguity Fixing Rate (With basis Recovery) [%]	Mean Ratio(Without Basis Recovery)	Mean Ratio (With Basis Recovery)	Proportion of Ratio Increase [%]
Single GPS	83.43	84.31	13.18	14.25	8.1
GPS/Galileo	89.25	89.75	8.33	8.83	6.0

As shown in Table 6, upon implementing the basis recovery method, the ambiguity fixing rate for the single GPS system rose by 0.9%, from 83.4% to 84.3%. Meanwhile, the fixing rate for the dual system experienced an increase of 0.5%, climbing from 89.25% to 89.75%. Furthermore, the mean ratio witnessed an improvement of 8.1% and 6.0% for the single GPS and dual system cases, respectively. The ambiguity fixing results, which are derived from tens of thousands of epochs over the full day, serve as compelling evidence of the proposed method's efficacy. Additionally, it is reiterated that the experiment was based on observation data collected in an open environment. This setup showcases the minimum level of gains that the method can offer. In obstructed environments, ambiguity fixing is expected to bring about even more significant improvements.

3.3. Experiment on the Recommended Number of Reference Stations

In this section, in an observation environment characterized by relatively flat terrain and inactive atmospheric conditions, four reference station networks of varying scales were carefully chosen. Then, an in-depth study was conducted to determine the recommended number of reference stations for users positioned within each of these networks. In the investigation of the number of reference stations, the basis recovery algorithm was employed. To illustrate its effectiveness, the proportion of recovered corrections relative to the total number of corrections is calculated and presented for different networks.

The four reference station networks described in the dataset section are used to interpolate the ionospheric corrections for the user station. By regarding the ionospheric delays derived at the user stations as precise references, the errors of the interpolated ionospheric corrections are calculated. Subsequently, these errors are used to calculate the root mean square (RMS) statistical results. To emphasize again, when selecting surrounding reference stations for this experiment, we tried to ensure they were evenly distributed in all directions. The results are shown in Figures 7–10. The reference station densities in Networks A and B are relatively dense, with inter-station distances predominantly clustered around 30 km and 50 km, respectively. As shown in Figures 7 and 8, the interpolation error of the L1 ionospheric corrections decreases significantly with an average of four reference stations, while additional reference stations lead to only marginal improvements in the error reduction. Ionospheric delay corrections involve both code and phase ionospheric corrections for all observed satellites. However, with the addition of each reference station, the communication load increases, resulting in an inefficient use of resources. In small-scale networks with station spacings of 50 km or less, it is recommended to use four or five reference stations. For networks with a station spacing of 100 km, an increase in the number of stations to six or seven still results in a measurable improvement; however, the benefit becomes negligible when the station count reaches eight. For large-scale networks spanning over 100 km, the use of six or seven reference stations is recommended to provide correction interpolations.

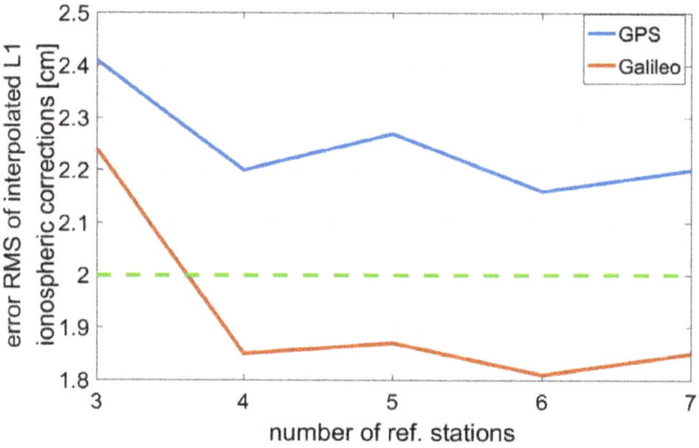

Figure 7. RMS values of the L1 ionospheric correction interpolation errors for Network A with different numbers of stations.

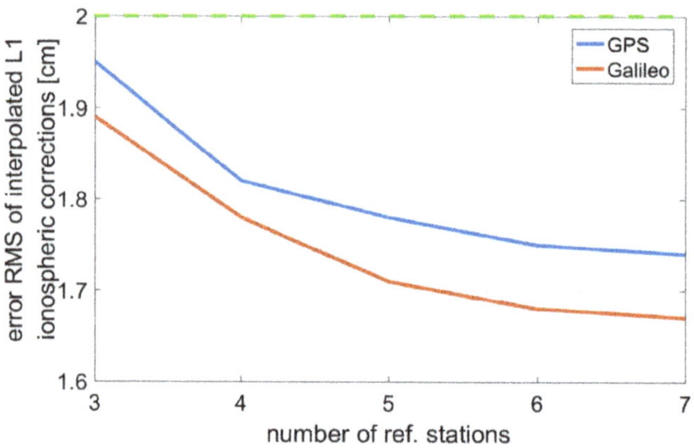

Figure 8. RMS values of the L1 ionospheric correction interpolation errors for Network B with different numbers of stations.

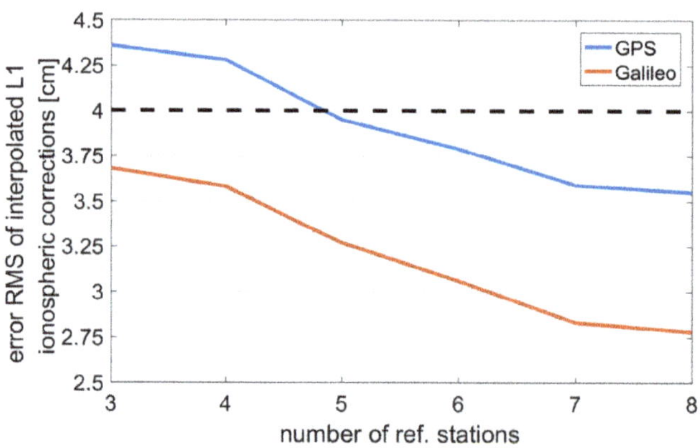

Figure 9. RMS values of the L1 ionospheric correction interpolation errors for Network C with different numbers of stations.

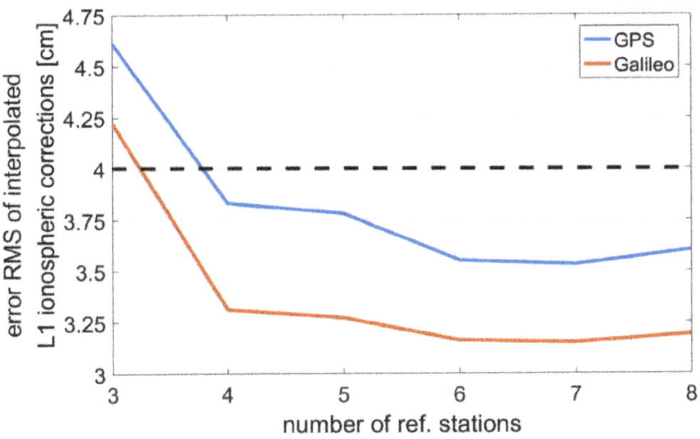

Figure 10. RMS values of the L1 ionospheric correction interpolation errors for Network D with different numbers of stations.

As the number of reference stations utilized for interpolation grows, the quantity of corrections employed in the interpolation process also increases. This heightens the probability of a basis deficiency for these corrections. To conduct a more in-depth assessment of the performance of the proposed method, Table 7 displays the number of corrections used for basis recovery in reference station networks of varying scales, along with their proportion relative to the total number of corrections. It can be observed that, for large networks with 100 km station spacing, when the number of reference stations reaches seven, the interpolation basis recovery algorithm restores anomalous values, preserving approximately 3% of the total observations. The method effectively utilizes observations that would otherwise be discarded, improving the efficiency of data utilization. It must be reiterated here that the observational data employed in the experiment were obtained from a network of reference stations that were evenly distributed in a flat terrain during a period of low ionospheric activity, and a low-order surface interpolation model was used. Consequently, the experimental results are applicable to this most prevalent and generalized scenario. However, it should be noted that the outcomes may vary when the observation environment or the distribution of the reference stations changes. Additionally, we must acknowledge that while increasing the number of reference stations can enhance the accuracy of interpolation corrections, as suggested by our recommendations, users may sometimes find it impractical to access such a large number of nearby reference stations. For instance, there might only be three to five reference stations available within a certain range, as discussed in Section 3.2. Moreover, regarding the influence caused by interpolation correction errors of varying magnitudes on PPP-RTK results, one can consult Wang et al. [23].

Table 7. Number (Num.) of recovered corrections and the proportion (Prop.) (%) of these corrections relative to the total number of corrections for different station networks with varying reference station densities.

Number of Reference Stations	Network A		Network B		Network C		Network D	
	Num.	Prop.	Num.	Prop.	Num.	Prop.	Num.	Prop.
4	1181	1.8	1383	2.0	1406	2.0	1419	2.1
5	1273	1.9	1535	2.2	1651	2.4	1312	1.9
6	1229	1.9	1939	2.8	1726	2.5	1548	2.3
7	1318	2.0	2114	3.1	1911	2.8	1964	2.9
8	-	-	-	-	2252	3.3	2033	3.0

4. Discussion

With the aim of engaging in a more in-depth discussion of the experimental results and showcasing their practical engineering significance, two auxiliary lines are added in Figures 7–10: one in green (representing a 2 cm threshold) and the other in black (representing a 4 cm threshold). These two thresholds were determined based on empirical evidence. When the mean RMS of the corrections is less than 2 cm, the system can generally tolerate this level of error, having a minimal impact on the estimation results. Therefore, the corrections at this error level can be regarded as accurate, and users can directly employ the ionosphere-fixed model for calculations. When the RMS level falls between 2 cm and 4 cm, using the ionosphere-fixed model results directly may be influenced to a certain extent, with the degree of influence being related to factors such as the number of satellites and the satellite geometric configuration. In such cases, users can choose to use either the ionosphere-fixed or the ionosphere-constrained estimation model based on the observation situation. When the mean RMS exceeds 4 cm, the solution will be significantly affected, potentially leading to consequences such as a marked increase in parameter errors or failure in ambiguity resolution. In this scenario, it is recommended to adopt the ionosphere estimation model.

When users are in the process of selecting reference stations for interpolating corrections, the first step is to examine the quantity of correction numbers offered by each reference station. Owing to issues like observation data quality and IAR, the number of corrections from certain individual reference stations may be significantly lower than that of their stations. These potentially unreliable reference stations must be excluded during the selection. Next, users should strive to select a set of reference stations that are not only close but also evenly distributed. This arrangement is crucial for ensuring that the quality of the interpolated corrections is optimized. Finally, users should aim to pick reference stations in a number that aligns with the recommended quantity specified in this paper. In summary, on the premise of ensuring that the quality and quantity of the corrections provided by the selected reference stations remain normal, among the group of reference stations nearest to the user (which implies that the distances between these reference stations and the user should not differ significantly), one should strive to select the recommended number of reference stations in different azimuths relative to the user. In the case of a sparse station network, it is necessary to choose at least three reference stations, ensuring that the user is located within the Delaunay triangle formed by these three stations.

5. Conclusions

Interpolating ionospheric delay corrections at the user station, extracted from reference stations, plays a crucial role in PPP-RTK applications. In non-specific environments, the quality of interpolated corrections depends not only on the choice of the interpolation model but also on the reference stations used in the interpolation process. Furthermore, determining the number of reference stations based on the size of the reference network and mitigating correction losses from individual stations that result in the unavailability of interpolated corrections and consequently lead to the wastage of observations, are two crucial factors to address.

This study models and analyzes two common scenarios of correction loss, with particular emphasis on the situation where the loss of corrections for certain satellites at a single reference station causes discrepancies between the interpolated corrections and those of other satellites, leading to the exclusion of observations. To address this problem, the basis recovery algorithm is proposed. The algorithm effectively extracts and compensates for the combined receiver hardware delay in ionospheric delay interpolation. With this method, even if the corrections for a satellite are lost at a specific reference station, they can still be interpolated and compensated, ensuring the observation remains usable. The experimental results validated the feasibility of the algorithm, showing that the model interpolation errors are minimal, with the majority confined to within 1 cm. Furthermore, the linearly combined receiver hardware delays exhibit a consistent pattern across the satellites. The recovered ionospheric delay corrections obtained through the algorithm deviate from the accurate interpolated (calculated using a sufficient number of reference stations) values by no more than ±1 cm. This level of accuracy is adequate for PPP-RTK users. Furthermore, approximately 3% of the observations, which would otherwise have been discarded due to the missing corrections from a specific reference station, are retained by the algorithm.

Owing to the limitations in the accuracy of the interpolation model, increasing the number of reference stations beyond a certain threshold does not improve the interpolation precision; rather, it leads to a higher communication load. The threshold also depends on the size of the reference network. In this study, in the case of flat terrain, stable atmospheric conditions, and relatively uniform distribution of reference stations in all directions, the selection of the number of reference stations was analyzed across four reference networks when using the low-order surface interpolation model, each varying in scale and station density. Here, it is once again emphasized that the experimental conclusions are directly linked to the observational environment. The background circumstances assumed in this work represent the most general and widely applicable scenarios. In networks with station spacings of 30 km and 50 km, a notable reduction in interpolation error is observed when the number of reference stations reaches four, while additional increases in the number of stations result in only marginal improvements. For small networks with a station spacing of 50 km, the use of four or five reference stations is recommended. In contrast, for larger networks with a station spacing of 100 km, it is advisable to employ six or seven reference stations.

In conclusion, it is essential to highlight that the research in this work regarding the recommended number of reference stations focuses on ionospheric delay. With regard to the subject matter pertaining to tropospheric delay, it warrants further exploration and discussion.

Author Contributions: Conceptualization, R.T.; Methodology, S.W. and R.Z.; Software, S.W.; Validation, L.F. and X.L.; Writing—original draft, R.Z.; Supervision, R.T. and L.F.; Project administration, X.L. All authors have read and agreed to the published version of the manuscript.

Funding: This research was funded by the National Science Foundation for Young Scientists of China (Grant No. 42104020), the National Natural Science Foundation of China (Grant No. 12203060), and the Youth Innovation Promotion Association CAS (Grant No. 2023426).

Data Availability Statement: Some or all data, models, or code generated or used during the study are available from the corresponding author by request.

Acknowledgments: The authors would like to thank the IGS for supporting precise products. The authors are grateful to the American National Oceanic and Atmospheric Administration (NOAA) for the CORS data sharing platform.

Conflicts of Interest: The authors declare no conflict of interest regarding the publication of this paper.

References

1. Héroux, P.; Kouba, J. GPS precise point positioning using IGS orbit products. *Phys. Chem. Earth Part A-Solid Earth Geod.* **2001**, *26*, 573–578. [CrossRef]
2. Zumberge, J.F.; Heflin, M.B.; Jefferson, D.C.; Watkins, M.M.; Webb, F.H. Precise point positioning for the efficient and robust analysis of GPS data from large networks. *J. Geophys. Res. Solid Earth* **1997**, *102*, 5005–5017. [CrossRef]

3. Geng, J.; Teferle, F.N.; Meng, X.; Dodson, A.H. Towards PPP-RTK: Ambiguity resolution in real-time precise point positioning. *Adv. Space Res.* **2011**, *47*, 1664–1673. [CrossRef]
4. Laurichesse, D.; Mercier, F.; Berthias, J.P.; Broca, P.; Cerri, L. Integer ambiguity resolution on undifferenced GPS phase measurements and its application to PPP and satellite precise orbit determination. *Navigation* **2009**, *56*, 135–149. [CrossRef]
5. Collins, P.; Lahaye, F.; Heroux, P.; Bisnath, S. Precise Point Positioning with Ambiguity Resolution using the Decoupled Clock Model. In Proceedings of the 21st International Technical Meeting of the Satellite Division of The Institute of Navigation (ION GNSS 2008), Savannah, GE, USA, 16–19 September 2008.
6. Ge, M.; Gendt, G.; Rothacher, M.; Shi, C.; Liu, J. Resolution of GPS carrier-phase ambiguities in Precise Point Positioning (PPP) with daily observations. *J. Geod.* **2008**, *82*, 389–399. [CrossRef]
7. Geng, J.; Meng, X.; Dodson, A.H.; Teferle, F.N. Integer ambiguity resolution in precise point positioning: Method comparison. *J. Geod.* **2010**, *84*, 569–581. [CrossRef]
8. Wübbena, G.; Bagge, A.; Schmitz, M. RTK Networks based on Geo++® GNSMART- Concepts, Implementation, Results. In Proceedings of the 14th International Technical Meeting of the Satellite Division of The Institute of Navigation (ION GPS 2001), Salt Lake City, UL, USA, 11–14 September 2001.
9. Wübbena, G.; Schmitz, M.; Bagge, A. PPP-RTK: Precise Point Positioning Using State-Space Representation in RTK Networks. In Proceedings of the 18th International Technical Meeting of the Satellite Division of The Institute of Navigation (ION GNSS 2005), Long Beach, CA, USA, 13–16 September 2005.
10. Li, X.; Huang, J.; Li, X.; Lyu, H.; Wang, B.; Xiong, Y.; Xie, W. Multi-constellation GNSS PPP instantaneous ambiguity resolution with precise atmospheric corrections augmentation. *GPS Solut.* **2021**, *25*, 107. [CrossRef]
11. Li, X.; Zhang, X.; Ge, M. Regional reference network augmented precise point positioning for instantaneous ambiguity resolution. *J. Geod.* **2011**, *85*, 151–158. [CrossRef]
12. Odijk, D.; Zhang, B.; Khodabandeh, A.; Odolinski, R.; Teunissen, P.J.G. On the estimability of parameters in undifferenced, uncombined GNSS network and PPP-RTK user models by means of S-system theory. *J. Geod.* **2016**, *90*, 15–44. [CrossRef]
13. Teunissen, P.J.G.; Odijk, D.; Zhang, B. PPP-RTK: Results of CORS network-based PPP with integer ambiguity resolution. *J. Aeronaut. Astronaut. Aviat. Ser. A* **2010**, *42*, 223–230.
14. Zhang, B.; Teunissen, P.J.G.; Odijk, D. A Novel Un-differenced PPP-RTK Concept. *J. Navig.* **2011**, *64*, S180–S191. [CrossRef]
15. Amalfitano, D.; Cutugno, M.; Robustelli, U.; Pugliano, G. Designing and Testing an IoT low-cost PPP-RTK augmented GNSS Location device. *Sensors* **2024**, *24*, 646. [CrossRef] [PubMed]
16. Robustelli, U.; Cutugno, M.; Pugliano, G. Low-Cost GNSS and PPP-RTK: Investigating the Capabilities of the u-blox ZED-F9P Module. *Sensors* **2023**, *23*, 6074. [CrossRef]
17. Montenbruck, O.; Steigenberger, P.; Hauschild, A. Broadcast versus precise ephemerides: A multi-GNSS perspective. *GPS Solut.* **2015**, *19*, 321–333. [CrossRef]
18. Gu, S.; Gan, C.; He, C.; Lyu, H.; Hernandez-Pajares, M.; Lou, Y.; Geng, J.; Zhao, Q. Quasi-4-dimension ionospheric modeling and its application in PPP. *Satell. Navig.* **2022**, *3*, 24. [CrossRef]
19. Odolinski, R.; Teunissen, P.J.G. An assessment of smartphone and low-cost multi-GNSS single-frequency RTK positioning for low, medium and high ionospheric disturbance periods. *J. Geod.* **2019**, *93*, 701–722. [CrossRef]
20. Gao, Y.; Li, Z.; McLellan, J.F. Carrier Phase Based Regional Area Differential GPS for Decimeter-Level Positioning and Navigation. In Proceedings of the 10th International Technical Meeting of the Satellite Division of The Institute of Navigation (ION GPS 1997), Kansas City, MI, USA, 16–19 September 1997.
21. Han, S.; Rizos, C. GPS Network design and error mitigation for real-time continuous array monitoring systems. In Proceedings of the 9th International Technical Meeting of the Satellite Division of The Institute of Navigation (ION GPS 1996), Kansas City, MI, USA, 17–20 September 1996.
22. Wübbena, G.; Bagge, A.; Seeber, G.; Böder, V.; Hankemeier, P. Reducing distance dependent errors for real-time precise DGPS applications by establishing reference station networks. In Proceedings of the 9th International Technical Meeting of the Satellite Division of The Institute of Navigation (ION GPS 1996), Kansas City, MI, USA, 17–20 September 1996.
23. Wang, S.; Li, B.; Gao, Y.; Gao, Y.; Guo, H. A comprehensive assessment of interpolation methods for regional augmented PPP using reference networks with different scales and terrains. *Measurement* **2020**, *150*, 107067. [CrossRef]
24. Wang, M.; Gao, Y. An Investigation on GPS Receiver Initial Phase Bias and Its Determination. In Proceedings of the 2007 National Technical Meeting of The Institute of Navigation, San Diego, CA, USA, 22–24 January 2007.
25. Wang, S.; Li, B.; Ge, H.; Zhang, Z. Algorithm and Assessment of Ambiguity-Fixed PPP with BeiDou Observations and Regional Network Augmentation. *J. Surv. Eng.* **2020**, *146*, 04020009. [CrossRef]
26. Rovera, G.D.; Torre, J.M.; Sherwood, R.; Abgrall, M.; Courde, C.; Laas-Bourez, M.; Uhrich, P. Link calibration against receiver calibration: An assessment of GPS time transfer uncertainties. *Metrologia* **2014**, *51*, 476–490. [CrossRef]
27. Odijk, D. Fast Precise GPS Positioning in the Presence of Ionospheric Delays. Ph.D. Thesis, Delft University of Technology, Delft, The Netherlands, 2002.

28. Khodabandeh, A. Single-station PPP-RTK: Correction latency and ambiguity resolution performance. *J. Geod.* **2011**, *95*, 42. [CrossRef]
29. Melbourne, W.G. The case for ranging in GPS-based geodetic systems. In Proceedings of the First International Symposium on Precise Positioning with the Global Positioning System: Positioning with GPS—1985, Rockville, MA, USA, 15–19 April 1985; U.S. Department of Commerce: Rockville, MA, USA, 1985; pp. 373–386.
30. Wübbena, G. Software developments for geodetic positioning with GPS using TI-4100 code and carrier measurements. In Proceedings of the First International Symposium on Precise Positioning with the Global Positioning System: Positioning with GPS—1985, Rockville, MA, USA, 15–19 April 1985; U.S. Department of Commerce: Rockville, MA, USA, 1985; pp. 403–412.
31. Li, X.; Li, X.; Yuan, Y.; Zhang, K.; Zhang, X.; Wickert, J. Multi-GNSS phase delay estimation and PPP ambiguity resolution: GPS, BDS, GLONASS, Galileo. *J. Geod.* **2018**, *92*, 579–608. [CrossRef]
32. Wang, S.; Li, B.; Li, X.; Zang, N. Performance analysis of PPP ambiguity resolution with UPD products estimated from different scales of reference station networks. *Adv. Space Res.* **2018**, *61*, 385–401. [CrossRef]

remote sensing

MDPI

Review

High Precision Navigation and Positioning for Multisource Sensors Based on Bibliometric and Contextual Analysis

Jiayi Wei [1,2], Min Song [1,2,*] and Yunbin Yuan [1,2]

1 Innovation Academy for Precision Measurement Science and Technology, Chinese Academy of Sciences, Wuhan 430077, China
2 College of Earth and Planetary Sciences, University of Chinese Academy of Sciences, Beijing 100049, China
* Correspondence: msong@apm.ac.cn

Abstract: With the increasing demand for high-precision positioning, integrated navigation technology has become a key approach to achieving accurate and reliable location tracking in modern intelligent mobile platforms. While previous studies have explored the application of various sensor combinations, there is still a lack of systematic analysis regarding the integration of the four major sensors: GNSS, INS, vision, and LiDAR. This study analyzes 5193 academic articles published between 2000 and 2024 in the Web of Science database, employing bibliometric analysis, network analysis, and content analysis to evaluate the development and application of these four sensors in integrated navigation systems. By reviewing the evolution of integrated navigation technology, the study examines four typical integration modes: GNSS/INS, INS/visual, GNSS/INS/visual, and GNSS/INS/visual/LiDAR, discussing their complementarity, fusion algorithm optimization, and emerging application scenarios. Despite significant progress in improving navigation accuracy and environmental adaptability, challenges persist in sensor cooperation and real-time processing capabilities in complex environments. The study concludes by summarizing existing research findings and identifying gaps, with future research focusing on optimizing multisensor fusion algorithms, enhancing system adaptability, improving error models, and enhancing sensor performance in adverse environmental conditions.

Keywords: GNSS; multisource fusion navigation; INS; visual; LiDAR; state estimation methods

1. Introduction

Since the beginning of the 21st century, significant breakthroughs have been made in multisensor technologies, including global navigation satellite systems (GNSS), inertial navigation systems (INS), visual, and LiDAR. GNSS offers global coverage but requires direct satellite visibility, INS bridges temporary signal outages through motion propagation, while visual and LiDAR sensors provide environmental perception critical for sustained operation in GNSS-denied environments such as urban canyons and indoor facilities. However, an individual navigation system is often constrained by its inherent limitations. Therefore, it is difficult to provide high-precision and reliable navigation services in complex environments and limited conditions. Subsequently, integrated navigation technology emerged. This combines multiple navigation systems to effectively overcome the shortcomings of individual systems [1]. It has been demonstrated that integrated navigation has significant adaptability and excellent performance, and has become an important research technology in modern navigation. The continuous development and application of integrated navigation technology has significantly broadened its application across various fields. Initially, integrated navigation primarily focused on the combination of GNSS and INS. However, with advancements in technology, an increasing number of sensors have been integrated into these systems, enabling reliable navigation services even in diverse environmental conditions. There are several advantages of integrated navigation technology. First, it has high adaptability in complex environments and observation-limited conditions. This adaptability is achieved through synergistic sensor interactions: INS maintains trajectory continuity during GNSS signal loss, visual sensors adaptively extract navigational features from dynamic surroundings, and LiDAR constructs persistent spatial references—collectively enabling reliable positioning under challenging conditions such as signal-blocked urban areas, feature-deprived tunnels, and rapidly changing operational scenes. Such multimodal coordination ensures reliable positioning and navigation in diverse applications ranging from high-altitude flight and deep-sea exploration to autonomous driving in densely populated urban areas and unmanned operations in challenging terrains. Even in worse observation conditions, such as GNSS signal obstruction, severe environmental changes, and complex dynamic conditions, reliable navigation information is maintained by using integrated navigation technology [2]. Secondly, integrated navigation technology enables the optimal navigation solution for specific application scenarios by flexibly selecting and integrating different types of sensors and employing either loosely coupled (LC) or tightly coupled (TC) information fusion strategies [2,3]. Integrated navigation technology can be applied in various areas and plays a key role in technological advancement and industrial innovation due to its high degree of flexibility and versatility. With the ongoing advancement of this technology and increasing market demand, integrated navigation technology is advancing in various aspects. From the hardware perspective, novel sensors such as miniaturized INS, high-precision visual sensors, and long-range LiDAR have significantly enhanced the sensing capabilities of integrated navigation systems [4,5]. From the algorithmic perspective, state estimation methods, such as filtering methods, graph optimization, and deep learning, were gradually incorporated into navigation frameworks to improve system performance. Moreover, customized integrated navigation solutions, such as GNSS/INS, INS/visual, GNSS/INS/visual, and GNSS/INS/visual/LiDAR, have been widely adopted and validated in specific application scenarios. This has further explored the applications of integrated navigation technology.

To rigorously investigate the development of integrated navigation technology, this paper applies bibliometric analysis, network analysis, and content analysis methods [6] to comprehensively examine the references, authorship, countries, institutions, keywords, and citation impacts. The objective is to reveal the underlying knowledge structure, distribution of research capabilities, collaboration networks, key research areas, and emerging trends within the field of integrated navigation. These research results can provide robust data support for understanding the current state and trends of integrated navigation technology. Furthermore, the results can help to address future challenges in integrated navigation technology.

2. Research Approach

2.1. Data Collection

In order to explore the developmental trajectory of the knowledge structure, research hotspots, and emerging trends of integrated navigation technology, this study employs the Web of Science Core Collection database as the primary platform for literature retrieval and analysis. This platform includes a large number of high-quality, peer-reviewed journal articles, covering a wide range of academic disciplines, and provides comprehensive and systematic literature resources for this study. Additionally, the citation analysis tools of

WOS effectively reveal the citation relationships among the literature, helping to identify the academic trajectory, research hotspots, and emerging trends in the field [7].

Considering the widespread application of integrated navigation technology, we design two sets of keywords to comprehensively cover the relevant literature in the research field of GNSS/INS/visual/LiDAR integrated navigation. The first set includes "Integrated navigation", "Multi-sensor Fusion", "combined technology", and "navigation system", which focus on the core concepts of system integration and performance optimization, ensuring that the literature related to the fundamental theories and methods of integrated navigation is covered. The second set includes terms such as "GNSS", "GPS", "Global Navigation Satellite System", "INS", "Inertial Navigation System", "IMU", "Inertial Measurement Unit", "Vision", "Visual", "Camera-based Navigation", "Visual SLAM (simultaneous localization and mapping)", "LiDAR", and "Laser Radar". These keywords were selected because integrated navigation systems involve a wide range of technologies and sensors. This set covers both foundational technologies like GNSS and INS as well as advanced methods such as SLAM and LiDAR. By including these terms, the search ensures the retrieval of the literature covering both traditional and cutting-edge technologies in integrated navigation, providing a comprehensive view of the field's research progress. The two sets of keywords are linked by the Boolean operator "AND", while keywords within each set are connected by "OR" to ensure comprehensive coverage of all the relevant literature. The final search expression is formulated as follows: TS = ((Integrated navigation OR Multi-sensor Fusion OR combined technology OR navigation system) AND (((GNSS OR GPS OR Global Navigation Satellite System) AND (INS OR Inertial Navigation System OR IMU OR Inertial Measurement Unit)) OR ((GNSS OR GPS OR Global Navigation Satellite System) AND (Vision OR Visual OR Camera-based Navigation OR Visual SLAM)) OR ((GNSS OR GPS OR Global Navigation Satellite System) AND (LiDAR OR Laser Radar)) OR ((INS OR Inertial Navigation System OR IMU OR Inertial Measurement Unit) AND (Vision OR Visual OR Camera-based Navigation OR Visual SLAM)) OR ((INS OR Inertial Navigation System OR IMU OR Inertial Measurement Unit) AND (LiDAR OR Laser Radar)) OR ((Vision OR Visual OR Camera-based Navigation OR Visual SLAM) AND (LiDAR OR Laser Radar)))) AND DT = (Article OR Review) AND LA = English. Using this search formula in the Web of Science, we initially retrieved 5986 relevant papers. Through rigorous selection to exclude non-navigation articles, we focused on 5193 research papers published during the period from 2000 to 2024 for further bibliometric and network analysis. In order to support the comprehensive study, key citation information was extracted, including author names, article titles, publication details, and citation counts along with abstracts and reference lists.

2.2. Bibliometric, Network, and Content Analysis

In this paper, we adopted an integrated approach combining bibliometric, network, and content analysis to explore the structure, dynamics, and core content of published research [8]. Bibliometric analysis as a quantitative method aims to reveal and emphasize the patterns and impact of scientific output.

Network analysis is an effective method for handling large-scale bibliometric data, which can utilize advanced computational tools to uncover interdisciplinary connections, construct academic network maps, and predict trends while identifying key research focuses. VOSviewer (version 1.6.20) and CiteSpace (version 6.3.R1) are two commonly used bibliometric analysis tools, each with their own unique features. VOSviewer is a visualization software developed by Leiden University in Leiden, the Netherlands and is widely favored for its flexibility and powerful functionality [8]. In this study, VOSviewer was used for network analysis to visually present the research findings in an intuitive graphical format. Content analysis was also applied to summarize key trends and develop a comprehensive framework for the literature review. Through bibliometric analysis, we present the rapidly developing research topics and potentially uncover new topics to deepen the understanding of the current status of the technology and to provide a guide for future research topics. CiteSpace, developed by Chen [9,10], focuses on trend analysis and pattern recognition. It is highly effective in detecting emerging topics, research frontiers, and the evolution of academic networks. Through citation analysis, CiteSpace reveals the core literature, helping researchers identify key knowledge and developmental trends within a specific field [6]. To provide a clear overview of the research methodology, Figure 1 illustrates the key steps involved in data collection, processing, and analysis, offering a comprehensive framework for understanding the research approach.

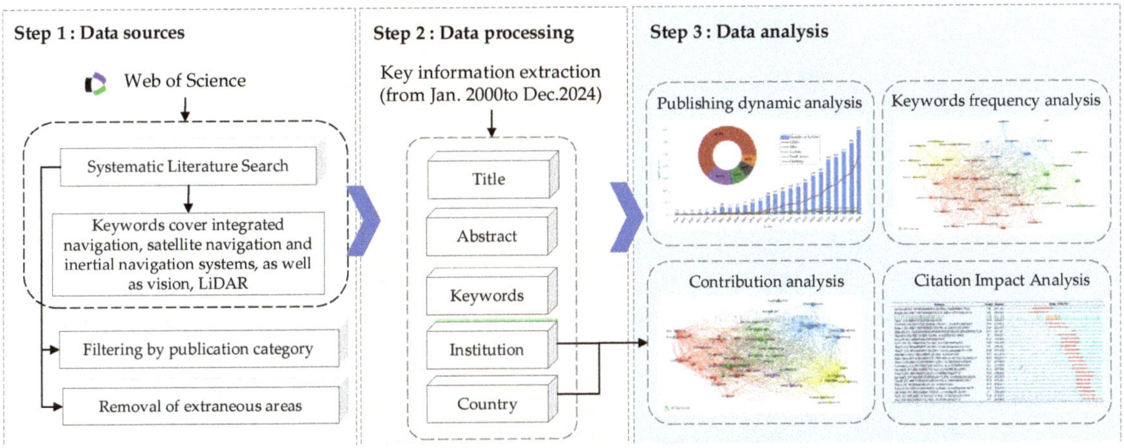

Figure 1. Flowchart of the research methodology for data collection, processing, and analysis.

3. Results

3.1. Analysis of Publication Outputs

The time series of publications on GNSS, INS, visual, and LiDAR-integrated navigation technology reveals changes in research intensity, the evolution of research focus, academic interest, and research and development investment. An upward trend in the total number of publications typically can identify an increase in academic attention or advances in core technologies that drive further research. The research of spatial distribution can be outlined as follows based on an analysis of 5193 publications from the Web of Science.

Figure 2 illustrates the geographical distribution of global scientific publications. It highlights the top five countries by number of publications, with China leading at 46.6%, followed by the USA at 12.4%, and Canada at 7.7%. South Korea and Germany rank fourth and fifth with 5.2% and 4.5%, respectively. These data not only reflect the relative weight of each country in global scientific publications but also reveal their competitiveness in terms of research investment, human resources, innovation capacity, and institutional efficiency. This helps us better understand the role and development trends of different countries in the research of GNSS/INS/visual/LiDAR integrated navigation technology and provides a quantitative reference for identifying leading countries.

Figure 3a illustrates the changes in the number of publications on integrated navigation research from 2000 to 2024. Based on the observed trends, the development of this field can be divided into three periods. The first period is from 2000 to 2005; the number of publications increased slowly, indicating that research in this field was relatively inactive and received little attention. The second period is from 2006 to 2011; the number of publications gradually increased, indicating increasing attention to integrated navigation technology. The third period is from 2012 to 2024; there was a rapid increase in publication numbers, especially in China. Figure 3b illustrates the total number of publications and citations from the top five countries between 2000 and 2024. China was far ahead in both metrics, indicating that it has paid more attention to the field of integrated navigation and has invested considerable effort in related research. With the development of multisensor fusion technology, the applications and research in integrated navigation are receiving more and more attention. Therefore, the number of publications has shown an almost exponential growth trend. This suggests the growing importance of the field. The research topics have also become more diversified, demonstrating that this field is still in a rapid development phase. Both new theories and technologies continually emerge.

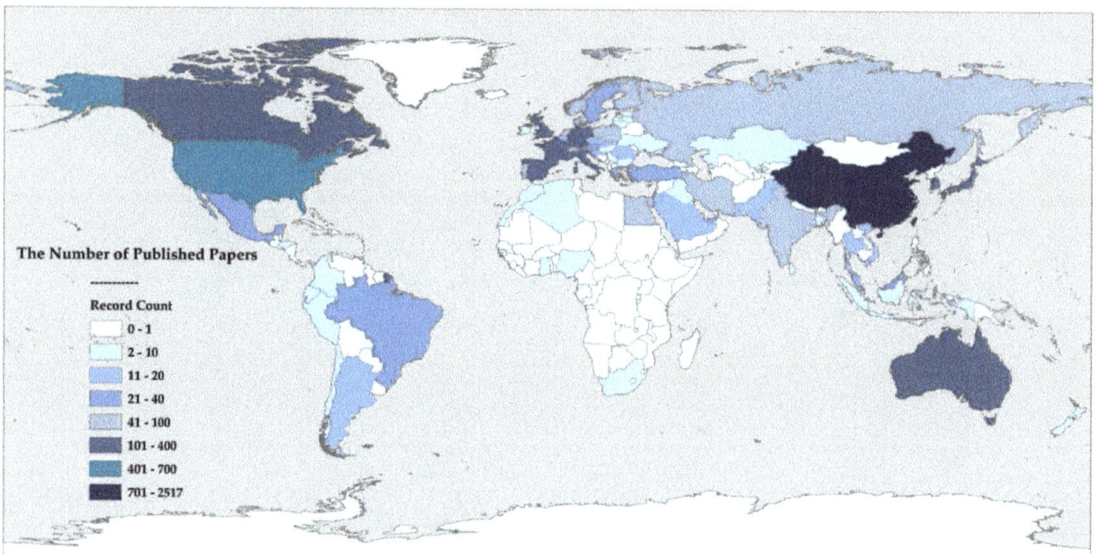

Figure 2. The number of global geographic distribution publications from 2000 to 2024.

(a) (b)

Figure 3. (**a**) The annual publications of countries in the field of integrated navigation from 2000 to 2024, with a donut chart showing the proportion of publications contributed by the top five countries; (**b**) total publications and citations of the top five countries in integrated navigation research from 2000 to 2024.

In terms of journal publications, there are 15 journals that have published more than 50 articles in this field. Table 1 lists the top 15 major journals based on the analytical sample of the published literature. These journals can be regarded as the most prominent publishing platforms in the field of integrated navigation. As shown in the table, *Sensors*, *IEEE Sensors Journal*, *Remote Sensing*, *IEEE Access*, and *IEEE Transactions on Instrumentation and Measurement* are the top five journals in this field. These data highlight that these journals play an important role in promoting academic exchange.

Table 1. Top 15 journals by the number of publications.

Rank	Journals	Publications
1	*Sensors*	493
2	*IEEE Sensors Journal*	226
3	*Remote Sensing*	209
4	*IEEE Access*	161
5	*IEEE Transactions on Instrumentation and Measurement*	138
6	*GPS Solutions*	114
7	*IEEE Transactions on Intelligent Transportation Systems*	104
8	*Journal of Navigation*	100
9	*IEEE Robotics and Automation Letters*	99
10	*Measurement Science and Technology*	86
11	*IEEE Transactions on Vehicular Technology*	76
12	*Measurement*	70
13	*Applied Sciences Basel*	68
14	*IEEE Transactions on Aerospace and Electronic Systems*	65
15	*Journal of Field Robotics*	55

3.2. Comprehensive Analysis of Authors, Countries, and Institutions

3.2.1. Author Contribution Analysis

Based on the VOSviewer analysis, Figure 4 shows a dense network of collaboration links with 533 core authors who have published at least 5 articles. Table 2 lists the top 15 authors ranked by the number of published articles in this field. Further analyses of Figure 4 and Table 2 reveal that most authors have significant advantages in both the number and strength of their collaborations, forming tight and frequent cooperative networks. This suggests that their broad collaborations play a crucial role in knowledge dissemination, interdisciplinary research, and scientific outputs.

Table 2. Top 15 ranking authors and number of articles published.

Rank	Author	Total Publications
1	Xiaoji Niu	84
2	Noureldin, Aboelmagd	65
3	El-Sheimy, Naser	49
4	Hsu, Li-Ta	48
5	Xingxing Li	37
6	Jingnan Liu	35
7	Chiang, Kai-Wei	35
8	Jian Wang	33
9	Quan Zhang	33
10	Shengyu Li	31
11	Tisheng Zhang	30
12	Yuxuan Zhou	29
13	Weisong Wen	29
14	Zhouzheng Gao	28
15	Hongping Zhang	28

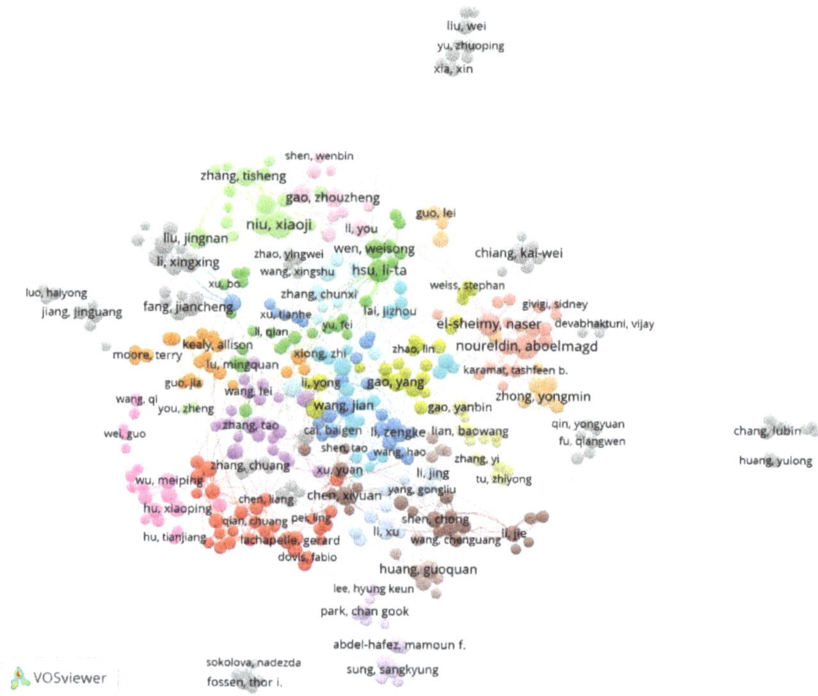

Figure 4. Knowledge map of author collaboration. Each node on the map represents an author, with the size of the node corresponding to the number of articles they have published. The connections between nodes illustrate collaboration between authors, with thicker lines indicating stronger and closer collaboration. The color of each node indicates the author's affiliation to a specific research cluster, with each color representing a distinct group of authors with stronger collaboration and related research topics.

Figure 5 illustrates the cocitation network of researchers who have made significant contributions to the multimodal fusion navigation technologies involving GNSS, INS, visual, and LiDAR. This network was generated through an in-depth exploration of the relevant literature database, where each dot represents an author who has published at least 15 papers and received at least 200 citations. A total of 56 core scholars were selected from 13,662 potential authors who met this strict criterion.

Table 3 lists the top 15 authors ranked by citation frequency in this cocitation network. Shaojie Sheng ranks first with 3314 citations and Xiaoji Niu with 2163 citations is close behind. They have demonstrated significant academic influence. The citing and recording of references possess the function of academic evaluation. It plays an important role in reflecting the publishing value of academic papers. A comparison between Tables 2 and 3 reveals that Xiaoji Niu (Wuhan University) and Noureldin, Aboelmagd (Queen's University/Royal Military College of Canada) is not only a leading author in publication count but also ranks highly in citation frequency. Additionally, scholars such as El-Sheimy, Naser (University of Calgary) and Hsu, Li-Ta (The Hong Kong Polytechnic University) have also produced substantial research results with high citation rates, clearly demonstrating their profound expertise and impact in GNSS/INS/visual/LiDAR integrated navigation technology.

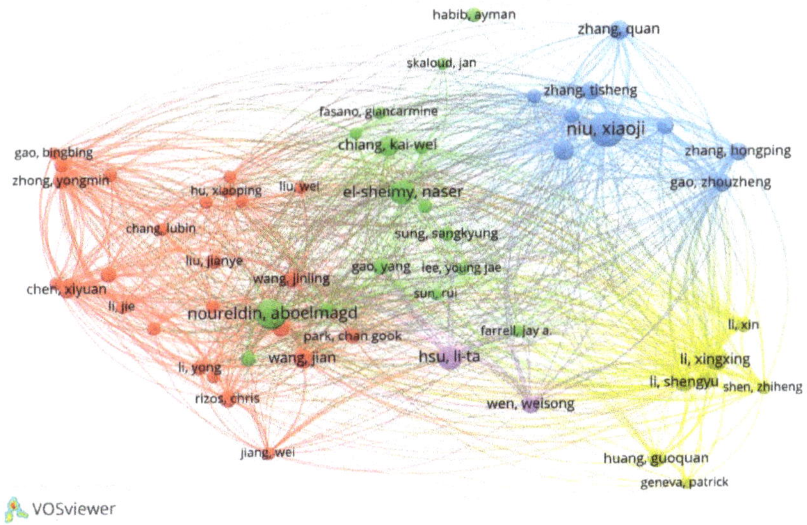

Figure 5. Knowledge map of cocitation authors. Nodes represent each author. Size of node defines the number of citations by each author. Link between nodes defines the cocitation relationship between authors. Width of link defines the strength of cocitation relationship. Colors define the cocitation-based cluster to which each author belongs.

Table 3. Top 15 ranking cocited authors and number of citations.

Rank	Author	Citations
1	Shaojie Sheng	3314
2	Xiaoji Niu	2163
3	Noureldin, Aboelmagd	2095
4	Scaramuzza, Davide	1621
5	El-Sheimy, Naser	1418
6	Yongmin Zhong	1264
7	Hsu, Li-Ta	1239
8	Stergios I. Roumeliotis	1218
9	Wei Xu	1104
10	Fu Zhang	1104
11	Anastasios I. Mourikis	1088
12	Gaoge Hu	1067
13	Guoquan Huang	910
14	Chang, Kai-Wei	863
15	Jinling Wang	860

3.2.2. Geographical Distribution of Contributions

To conduct an in-depth analysis of international research collaborations, a threshold of at least 15 coauthored publications was set. Among the 98 countries analyzed, 41 met this criterion and were included in the construction of the country collaboration network, as shown in Figure 6. Based on highly consistent collaboration patterns within the specified research field, the network is further divided into several distinct clusters [7]. The clusters of different colors reveal collaboration networks between different countries. For instance, the blue cluster highlights the intensive collaboration between countries such as China and the United States, demonstrating their significant joint contributions to research activities, while the green cluster emphasizes the close partnerships between countries like Canada, England, and India, showcasing their collaborative research efforts within the region. Each node in the figure represents a country and its corresponding number of published papers and the size of the node is proportional to the total number of academic articles published. Specifically, China is the largest node with 2422 papers published during the sample period, followed by the United States with 642 papers and Canada with 402 papers. Moreover, the thickness of the connections indicates the strength of collaboration between the associated countries, with the thicker lines indicating that the scientific collaboration between the two countries is stronger and more active. As shown in Figure 6, countries with a higher number of published papers tend to exhibit stronger collaborative relationships in the international cooperation network.

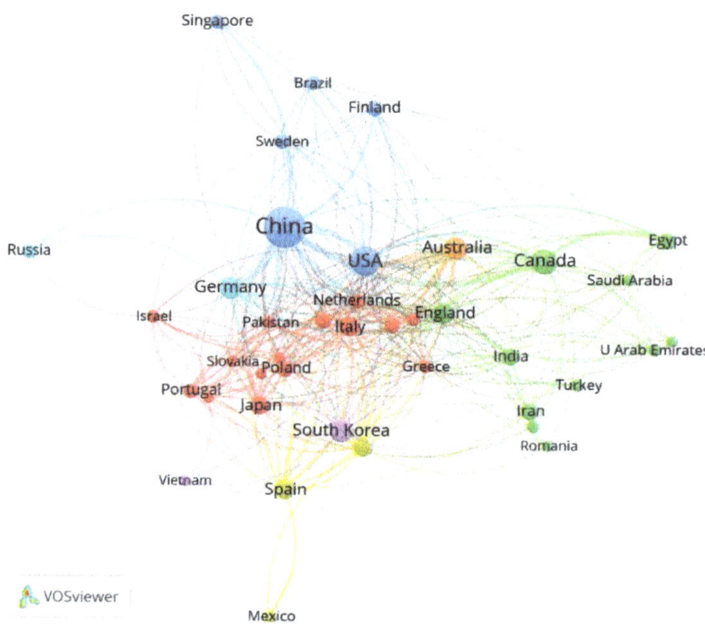

Figure 6. Knowledge map of country collaboration.

3.2.3. Institutional Collaboration Patterns

According to the available data, a total of 3108 different institutions have contributed to publications in the field of integrated navigation technology. As shown in Figure 7 and Table 4, among the top 15 institutions ranked by the number of publications, Wuhan University holds the top position, with 262 high-quality papers published in this field. This was followed by Beihang University with 199 related research result achievements. Additionally, institutions such as the University of Calgary, the National University of Defense Technology, and Southeast University have also gained leading positions due to their outstanding research performance. The contributions of these institutions are reflected not only in the absolute number of their research results but also in their substantial strength and extensive influence in the research, development, and theoretical exploration of integrated navigation technology. Their research results provide continuous momentum for the advancement of integrated navigation technology and lay a solid foundation for the future development of this field.

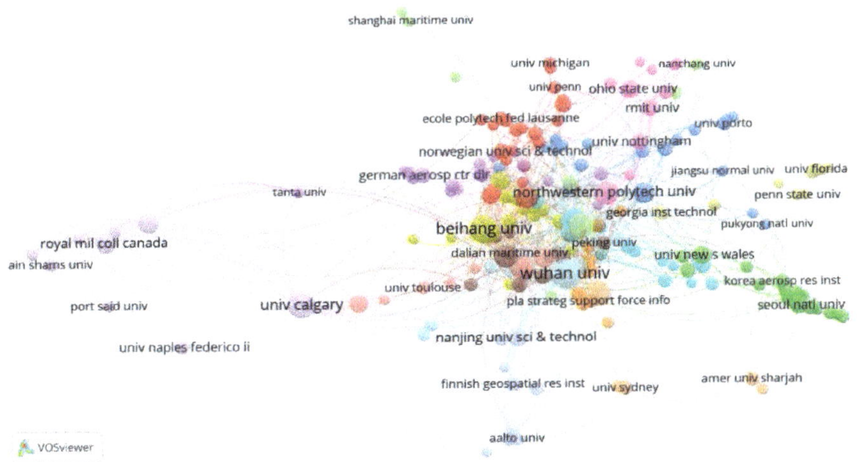

Figure 7. Knowledge map of institution collaboration.

Table 4. Top 15 ranking organizations and number of articles published.

Rank	Affiliations	Articles
1	Wuhan University	262
2	Beihang University	199
3	University of Calgary	157
4	National University of Defense Technology	126
5	Southeast University	116
6	Chinese Academy of Science	106
7	Northwestern Polytechnical University	103
8	Harbin Engineering University	98
9	Nanjing University of Aeronautics and Astronautics	92
10	The Hong Kong Polytechnic University	78
11	Tsinghua University	77
12	Beijing Institute of Technology	72
13	China University of Mining and Technology	71
14	Shanghai Jiao Tong University	71
15	Royal Military College of Canada	63

3.3. *Research Hotspot Analysis*

3.3.1. Keyword Analysis

Figure 8 illustrates the keyword network of publications spanning from 2000 to 2024, with a co-occurrence frequency of at least 50. To ensure focus on frequently appearing and influential keywords, a "mean-split" standardization method was applied to measure the importance and relative frequency of the keywords [11]. The size of each node corresponds to the frequency of the keywords' appearance in the literature. Keywords with a frequency above the mean have larger nodes, while those below the mean have smaller nodes. The strength of the associations between the keywords is represented by the thickness and distance of the connections; the higher the co-occurrence frequency, the thicker the connection.

The figure displays four main research themes, distinguished by color: the blue cluster focuses on the broad applications of GNSS and their derived technologies in positioning and navigation, such as BDS, GPS, precise point positioning (PPP), and real-time kinematic (RTK); the green cluster focuses on INS and sensor fusion, covering technologies like accelerometers, GPS, and Kalman filters, aiming to enhance navigation accuracy and state estimation through sensor data fusion; the yellow cluster focuses on visual perception and positioning optimization and it includes visual–inertial odometry (VIO), pose estimation, feature extraction, optimization algorithms, and visualization techniques to enhance location awareness and accuracy in complex environments; the red cluster centers on autonomous driving and related technologies and it encompasses multisensor fusion, SLAM, LiDAR, deep learning, autonomous navigation, and visual navigation, leveraging advanced sensing and processing methods to improve real-time positioning and navigation for autonomous vehicles.

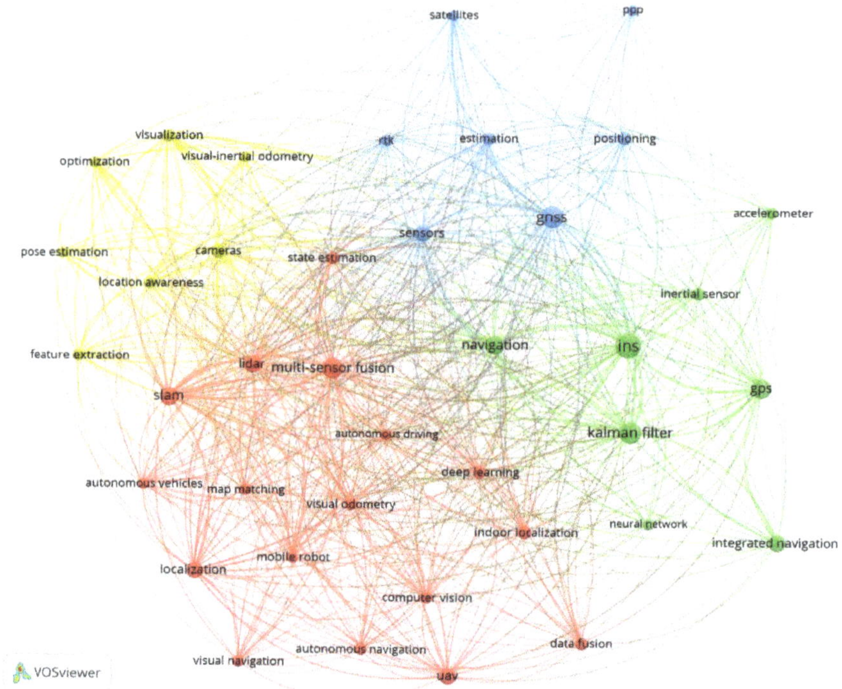

Figure 8. Visualization analysis of keyword co-occurrence network from 2000 to 2024.

In this study, we conducted an in-depth analysis of the 25 most notable keywords exhibiting sudden growth from 2000 to 2024 using the CiteSpace tool [12], as shown in Figure 9, to reveal the evolutionary trajectory of research themes. Each keyword is associated with specific year labels and intensity indicators, with red bar charts illustrating the periods of citation bursts and peak intensity. Through the temporal dynamic analysis of these keywords, we can clearly understand the changes in the popularity of research hotspots over time, thereby providing insights for future research directions.

Data from Figure 9 indicate that certain keywords highlight how early technological development primarily relied on the combination of INS and GPS. Between 2000 and 2008, inertial navigation displayed a burst intensity of 20.47, while the global positioning system reached 14.97 from 2000 to 2011. Together, these systems formed the foundational architecture of navigation systems, significantly enhancing their reliability. Meanwhile, GPS/INS integrated navigation, with a burst intensity of 23.97 from 2006 to 2016, emerged as a representative example of multimodal fusion, meeting the demand for high-accuracy positioning in complex scenarios [13,14]. Subsequently, computer vision achieved a burst intensity of 8.63 from 2012 to 2014 and vision technologies reached 7.65 from 2013 to 2016, highlighting a notable improvement in environmental perception capabilities and expanding the application scope of sensor combinations. Additionally, Figure 9 reveals profound transformations in the integration of theoretical innovation and practical application within integrated navigation algorithms, transitioning from the classic Kalman filter to intelligent developments, including particle filter, factor graph optimization, deep learning, machine learning, and the adaptive Kalman filter. In recent years, the emergence of "real-time systems", "location awareness", "pose estimation", and "ambiguity resolution" reflects notable advancements in integrated navigation. These terms highlight a growing focus on improving real-time responsiveness, enhancing environmental adaptability, and achieving greater positioning accuracy [15–20]. In summary, the developmental trajectory of these keywords not only revisits the key milestones in integrated navigation technology but also provides significant insights for future research trends, highlighting the broad application prospects of intelligent algorithms in navigation systems.

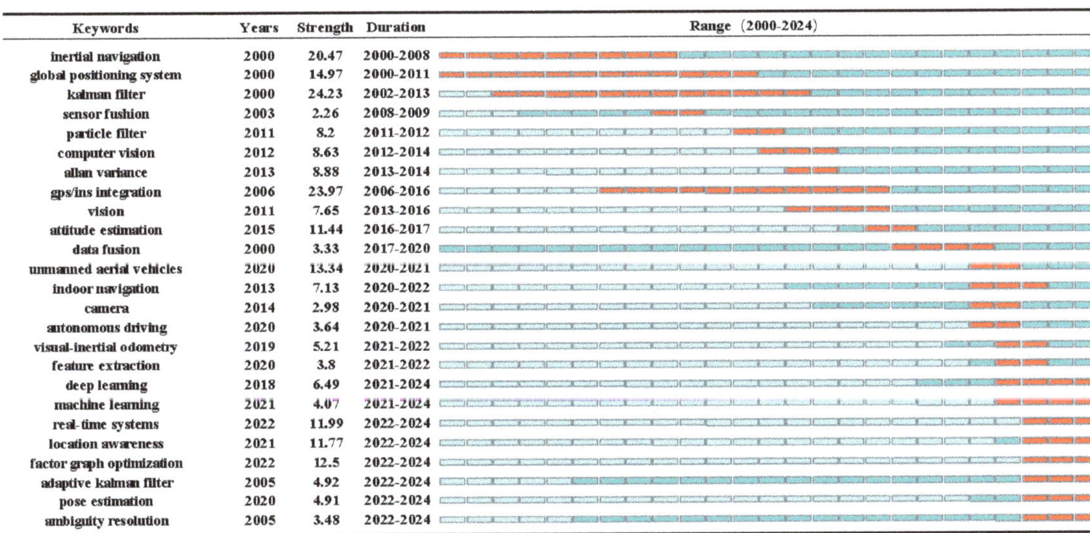

Figure 9. The top 25 burst keywords from 2000 to 2024. The dark blue band reflects the starting year of the keyword and the red band reflects the period of its sudden activity.

3.3.2. Citation Impact Analysis

Citation analysis serves as a powerful tool that reveals the cocitation relationships among various works, aiding researchers in exploring the intrinsic connections within the literature knowledge system. This analytical approach effectively identifies key works that have significantly influenced a specific research field and quantifies the strength of the interconnections among these works, thereby providing an effective assessment of the research's impact and significance in the academic community [21].

Highly cited papers often signify pioneering value within their respective research fields and play a crucial role in the development of the entire domain. Table 5 lists the top ten most cited works in the field of integrated navigation. These studies cover a wide range of topics, including traditional GNSS/INS integration, VIO, and LiDAR–inertial fusion, highlighting significant advancements in multisensor fusion, state estimation accuracy, and real-time performance optimization. For example, in document 10, Noureldin et al. focus on enhancing the performance of low-cost navigation systems by integrating MEMS-based INS and GPS. Their approach improves filtering algorithms to mitigate multipath effects and signal obstructions, significantly increasing navigation accuracy and reliability in urban environments [22]. Similarly, document 9 surveys various in-car positioning and navigation technologies, including GNSS, inertial sensors, and odometry, highlighting their practical applications in modern automotive systems [23]. In the domain of visual–inertial odometry, document 2, Qin et al.'s research on VINS-mono achieves high-precision state estimation by fusing data from visual and inertial sensors, making it particularly suitable for autonomous navigation systems in GPS-denied environments and advancing the development of drone and robotic navigation [24]. Document 5 further refines visual–inertial odometry through manifold preintegration techniques, enhancing real-time performance and accuracy and broadening applications in autonomous driving scenarios. LiDAR–inertial fusion is also a key focus [25]. In document 8, "FAST-LIO2", integrates LiDAR data with inertial measurements to implement a fast direct odometry approach. This innovation significantly improves real-time navigation efficiency and robustness in dynamic environments, paving the way for next-generation autonomous navigation systems [26]. Overall, these works illustrate a clear progression in integrated navigation technologies. From traditional GNSS/INS combinations to cutting-edge visual–inertial and LiDAR–inertial methods, they have collectively elevated the state of the art, enabling more accurate, reliable, and versatile navigation across a variety of challenging operational environments.

Table 5. Top ten articles of cocitation literature [22–31].

Rank	Title	Author	Year
1	Vision Meets Robotics: The KITTI dataset	Geiger, Andreas	2013
2	VINS-Mono: A Robust and Versatile Monocular Visual-Inertial State Estimator	Tong Qin	2018
3	ORB-SLAM3: An Accurate Open-Source Library for Visual, Visual-Inertial, and Multimap SLAM	Campos, Carlos	2021
4	AUV Navigation and Localization: A Review	Paull, Liam	2014
5	On-Manifold Pre-integration for Real-Time Visual–Inertial Odometry	Forster, Christian	2017
6	High-precision, consistent EKF-based visual-inertial odometry	Mingyang Li	2013
7	Visual-Inertial Monocular SLAM With Map Reuse	Mur-Artal, R	2017
8	FAST-LIO2: Fast Direct LiDAR-Inertial Odometry	Wei Xu	2022
9	In-Car Positioning and Navigation Technologies-A Survey	Skog, Isaac	2009
10	Performance Enhancement of MEMS-Based INS/GPS Integration for Low-Cost Navigation Applications	Noureldin, Aboelmagd	2009

The burst index is a metric that measures sudden citation surges, highlighting works that rapidly gain influence within a short timeframe. Such bursts often reflect a temporary but significant impact on research trends. In CiteSpace II, the burst index helps identify pivotal articles that signal rising importance or shifts in intellectual focus [32]. Figure 10 shows the top 25 documents ranked by their burst index, illustrating these abrupt spikes in citation activity. This literature highlights distinctive research advancements in sensor combination methods, algorithm optimization strategies, and multisensor fusion technologies. Within the field of multisensor fusion, the exploration of various combination approaches has consistently been a central focus. From early GNSS/INS integration to the rise of vision–inertial SLAM and the inclusion of LiDAR with inertial sensors, each added sensor has significantly contributed to enhancing navigation accuracy and robustness. For instance, the study by Groves PD demonstrated a burst index of 42.27 between 2014 and 2018, indicating significant progress in addressing signal occlusion in complex environments through GNSS and inertial navigation system integration [33]. The vision–inertial SLAM algorithm proposed by Leutenegger S had a burst index of 52.37 during the 2017 to 2020 period, showcasing the maturity of this sensor combination in resolving localization drift problems in dynamic environments [34]. Subsequently, the study by Qin T incorporated

LiDAR into a multimodal sensor fusion system, achieving a burst index of 55.9 during the 2020 to 2024 period, a milestone that greatly improved the performance of unmanned and robotic navigation technologies [24].

References	Strength	Duration	Range（2020-2024）
Jazwinski AH, 2007, DOVER BOOKS ON ELECTRICAL ENGINEERING, V0, P0	3.88	2007-2011	
Hong SP, 2005, IEEE T VEH TECHNOL, V54, P731, DOI 10.1109/TVT.2004.841540	4.68	2006-2010	
Groves PD, 2008, ARTECH HSE GNSS TECH, V0, P3	23.89	2010-2013	
Farrell J, 2008, AIDED NAVIGATION GPS, V0, P0	22.59	2010-2013	
Noureldin A, 2009, IEEE T VEH TECHNOL, V58, P1077, 10.1109/TVT.2008.926076	22.49	2010-2014	
Georgy J, 2010, IEEE T VEH TECHNOL, V59, P599, 10.1109/TVT.2009.2034267	10.44	2011-2015	
Kaplan ED, 2017, UNDERSTANDING GPS/GNSS PRINCIPLES AND APPLICATIONS, V0, P0	26.57	2017-2017	
Martinelli A, 2012, IEEE T ROBOT, V28, P44, 10.1109/TRO.2011.2160468	14.7	2013-2017	
Groves PD, 2013, ARTECH HSE GNSS TECH, VO, P1	42.27	2014-2018	
Li MY, 2013, INT J ROBOT RES, V32, P690, 10.1177/0278364913481251	22.62	2014-2018	
Zhang T, 2012, MEASUREMENT, V45, P691, 10.1016/j.measurement.2011.12.021	9.05	2014-2017	
Chen XY, 2013, MEASUREMENT, V46, P3847, 10.1016/j.measurement.2013.07.016	14.26	2015-2018	
Chowdhary G, 2013, J FIELD ROBOT, V30, P415, 10.1002/rob.21454	11.97	2015-2018	
Engel J, 2014, LECT NOTES COMPUT SC, V8690, P834, 10.1007/978-3-319-10605-2_54	22.43	2016-2019	
Forster C, 2014, IEEE INT CONF ROBOT, V0, PP15, 10.1109/ICRA.2014.6906584	21.9	2016-2019	
Leutenegger S, 2015, INT J ROBOT RES, V34, P314, 10.1177/0278364914554813	52.37	2017-2020	
Mur-Artal R, 2015, IEEE T ROBOT, V31, P1147, 10.1109/TRO.2015.2463671	38.21	2017-2020	
Forster C, 2017, IEEE T ROBOT, V33, P1, 10.1109/TRO.2016.2597321	34.52	2018-2022	
Mur-Artal R, 2017, IEEE ROBOT AUTOM LET, V2, P796, 10.1109/LRA.2017.2653359	21.56	2018-2022	
Yang ZF, 2017, IEEE T AUTOM SCI ENG, V14, P39, 10.1109/TASE.2016.2550621	14.78	2018-2021	
Falco G, 2017, SENSORS-BASEL, V17, P0, 10.3390/s17020255	14.78	2018-2021	
Usenko V, 2016, IEEE INT CONF ROBOT, V0, PP1885, 10.1109/ICRA.2016.7487335	12.18	2018-2021	
Mur-Artal R, 2017, IEEE T ROBOT, V33, P1255, 10.1109/TRO.2017.2705103	37.7	2019-2022	
Sun K, 2018, IEEE ROBOT AUTOM LET, V3, P965, 10.1109/LRA.2018.2793349	11.44	2019-2022	
Qin T, 2018, IEEE T ROBOT, V34, P1004, 10.1109/TRO.2018.2853729	55.9	2020-2024	

Figure 10. The top 25 references with the strongest citation bursts from 2000 to 2024. The dark blue band reflects the publication date of the paper and the red band reflects the period in which the paper was cited [22,24,25,30,31,33–52].

In the realm of algorithm optimization, research on sensor combinations has focused on enhancing the real time performance and accuracy of data fusion. The filtering algorithm proposed by Noureldin A achieved a burst index of 22.49 during the 2010 to 2014 period. This study significantly reduced the impact of multipath effects on navigation accuracy by improving filtering methods for low-cost inertial navigation systems [22]. The work of Mur-Artal R concentrated on improving visual–inertial SLAM algorithms, attaining burst indices of 38.21 and 37.7 during the 2017–2020 and 2019–2022 periods, respectively. These studies innovatively optimized loop closure detection algorithms and zero-drift localization methods, further enhancing the robustness and real-time performance of SLAM systems in dynamic environments, laying a critical algorithmic foundation for deeper multisensor fusion [35,36]. In conclusion, the evolution of sensor combination technology from single-sensor integration to multimodal fusion clearly reflects the development trajectory of navigation technologies. At each stage, the exploration of sensor combinations and algorithm optimization has consistently enhanced the accuracy and adaptability of navigation systems. From GNSS/INS to vision–inertial SLAM, and finally to multimodal fusion, navigation technology has been advancing toward a highly intelligent, integrated system. In the future, research will continue to improve accuracy, robustness, and applicability, providing reliable support for autonomous driving, robotic navigation, and other complex scenarios.

4. Discussion

4.1. Development and Innovation in Multisource Sensor Technologies

In the current era of information and intelligence, multisensor fusion has emerged as a pivotal technology for achieving high-precision positioning, navigation, environmental perception, and intelligent decision-making. By integrating data from various sensor types, multisensor fusion enables comprehensive and accurate environmental awareness, significantly enhancing system stability and reliability. However, each sensor type has its own strengths and limitations, which must be carefully considered to optimize the effectiveness of multisensor fusion in integrated navigation systems.

As illustrated in Figure 11, a visual analysis of relevant literature databases reveals a growing trend in multisensor fusion technology in recent years. GNSS provides accurate, large-area positioning and is widely used in outdoor applications, such as agricultural automation and open-road autonomous driving, due to its global coverage and high accuracy [53]. However, its reliability is compromised in urban canyons, dense forests, or tunnels due to signal obstruction, interference, and multipath effects. In contrast, INS, an autonomous and passive navigation system, operates independently of external signals and is less affected by environmental interference. It is widely adopted in missile guidance, underwater exploration, and aerospace navigation due to its robustness in extreme environments. However, its cumulative errors over time necessitate periodic calibration through external references (e.g., GNSS or visual landmarks) [54]. To compensate for the limitations of both GNSS and INS, GNSS/INS integration has become one of the most widely used techniques in integrated navigation systems [55]. The integration of GNSS and INS has evolved from independent, large-scale, and power-hungry systems to systems focusing on miniaturization, low power consumption, deep integration, and intelligence, thus laying a solid foundation for integrated navigation [54,56–62].

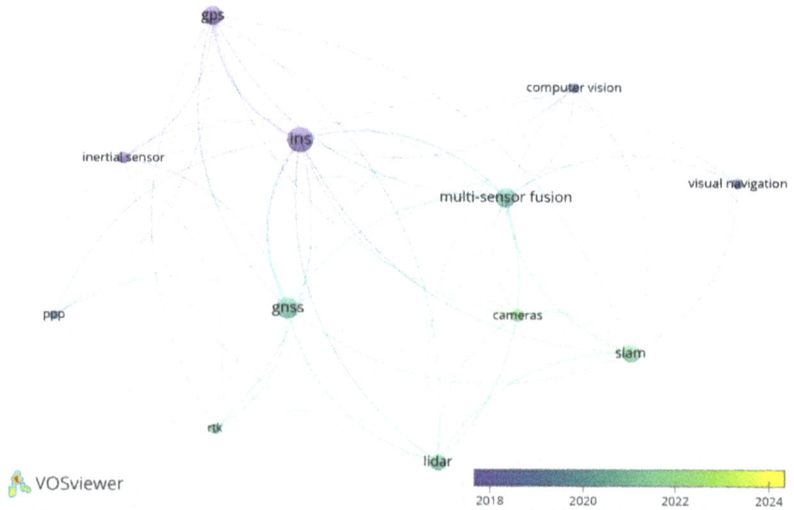

Figure 11. Connections and peak period of multisensor technologies in integrated navigation.

Around 2020, the incorporation of LiDAR technology marked a significant expansion in this field, greatly enhancing the environmental perception and three-dimensional mapping capabilities of navigation systems [63,64]. LiDAR provides high-precision 3D point cloud data, capable of capturing detailed terrain and obstacle information. It excels in applications requiring high-precision mapping, such as autonomous vehicles, topographic surveys, and infrastructure inspections. Unlike some other sensors, LiDAR is less affected by lighting conditions and can operate in both daylight and nighttime environments. However, LiDAR systems face challenges, including high cost, sensitivity to weather conditions (such as rain or fog), and substantial power consumption, limiting their large-scale deployment in certain applications [65]. Subsequently, around 2021, the rapid rise of visual sensor technology in integrated navigation applications became another significant milestone [66,67]. The introduction of advanced technologies such as SLAM significantly improved the robustness and accuracy of navigation solutions, especially in environments where GNSS signals are weak or unavailable, such as indoor settings or urban canyons. In these scenarios, visual sensors provide rich environmental information that compensates for the lack of GNSS signals. A practical example is their use in indoor warehouse robots, where visual sensors enable precise navigation along aisles and shelves without the need for external positioning infrastructure [68]. The integration of INS/visual and GNSS/INS/visual combinations effectively improves navigation capabilities. The INS/visual combination leverages visual data to compensate for INS drift, enhancing system stability and long-term accuracy. Meanwhile, the GNSS/INS/visual combination further improves navigation performance by integrating additional positioning information from both GNSS and visual sensors, thereby strengthening navigation capabilities in GNSS-denied or weak signal environments. However, the performance of visual sensors is highly dependent on environmental lighting conditions and affected by factors such as shadows, glare and low light, which increase the difficulty of real-time operation in complex and dynamic environments. For instance, in outdoor construction sites where lighting conditions change rapidly, visual sensors might struggle to maintain consistent performance, necessitating more robust processing algorithms or complementary sensors [69,70]. Nevertheless, the widespread application of visual sensors has accelerated the development of multisensor fusion systems, advancing navigation technologies toward more intelligent and flexible solutions, ultimately enabling breakthroughs in adaptive navigation systems. To further improve navigation accuracy and system stability, the integration of GNSS/INS/visual/LiDAR has emerged. The fusion of LiDAR point cloud data with GNSS and INS provides comprehensive sensing capabilities, further enhancing the system's navigation accuracy. However, this integration also introduces challenges related to computational complexity and real-time processing, which require continuous optimization to improve overall performance. In recent years, the integration of vision and language has emerged as an innovative solution, significantly enhancing the capabilities of navigation systems. Vision–language navigation combines visual input with natural language instructions, enabling more intuitive and flexible navigation solutions. For example, in robotics, autonomous vehicles, and aerial drones, systems can process both environmental images and verbal commands to navigate, improving autonomy in environments where GNSS signals are unavailable or traditional methods face challenges. The use of vision–language navigation in aerial drones, for instance, allows them to operate autonomously in complex environments by interpreting both visual and language cues, enabling them to make real-time decisions without relying on GNSS signals [71–73].

In summary, the progression from the classic integration of GNSS and INS to the incorporation of LiDAR and the application of visual sensors not only highlights the increasing maturity of multisensor information fusion technology but also demonstrates the vitality of ongoing innovation in the navigation field to meet diverse demands [74]. Looking ahead, future research will focus on addressing global, seamless navigation and positioning through low-cost integration methods as well as improving system miniaturization, real-time processing capabilities, and sensor fusion techniques. Additionally, there will likely be continued efforts to overcome the individual limitations of each sensor type, particularly in terms of weather sensitivity, power consumption, and environmental adaptability.

4.2. Integration and State Estimation in Multisource Systems

4.2.1. Integration Methods of Multisource Sensors

The integration methods of multisensor information play a crucial role in combination navigation systems, primarily categorized into LC and TC [50]. These two categories represent different levels of data integration and processing closeness. Based on the LC and TC methods, various sensor combinations are widely applied, including GNSS/INS, INS/visual, GNSS/INS/visual, and GNSS/INS/visual/LiDAR. Each combination leverages the strengths of different sensors to enhance navigation accuracy and reliability. In addition to the LC and TC approaches, GNSS/INS combinations have evolved to include deeply coupled (DL) and ultratightly coupled modes. These advanced methods achieve a higher level of integration by tightly coupling GNSS signal processing with INS data, significantly improving robustness and performance in challenging environments such as urban canyons or GNSS-denied areas. The following sections discuss the technical characteristics, advantages, and limitations of different integration methods along with their implementation in various multisensor configurations.

1. Loosely coupled integration mode

In an LC integration mode, each sensor operates independently, calculating its own positioning information, with data fusion occurring only during the postprocessing stage to enhance overall positioning performance. This fusion approach is straightforward, requiring no complex real-time interaction mechanisms. However, its positioning accuracy can be significantly affected by the performance fluctuations of individual sensors. In GNSS/INS systems when GNSS is unable to provide a navigation solution, the LC system will degrade to an INS mechanical alignment [55].

In the visual–inertial navigation system (VINS) employing an LC strategy, the visual system and the INS operate independently, each performing separate error estimation and correction. The visual system is primarily responsible for updating the position while the IMU continuously provides real-time dynamic information. During fusion, the two systems are connected through a simple association mechanism such as periodically applying external parameters from the visual system to correct IMU drift. This process does not involve real-time sharing and fusion of observational data.

In GNSS/INS/visual systems, the data processing and fusion of each sensor operate independently, without direct data interaction or constraints. For example, Neal A. Carlson [75] proposed a federated square-root filter design suitable for distributed parallel processing in multisensor systems. The overall operation is divided into multiple independent local subsets, each containing a local filter—specifically, GNSS/IMU and IMU/camera constitute two subfilters. The locally optimal estimates from these subfilters are then input into a main filter for solution, achieving significant performance improvement and optimization for real-time multisensor applications such as integrated navigation systems.

In GNSS/INS/visual/LiDAR systems, GNSS provides absolute position and time information, INS continuously calculates the vehicle's velocity, attitude, and position, the visual sensor obtains relative position or heading information through image processing, and LiDAR supplies distance and angle information from the environment. These data are processed individually and subsequently integrated to compensate for errors in the IMU's gyroscope and accelerometer measurements [76,77].

Overall, in a loosely coupled (LC) architecture, each sensor operates independently with separate processing pipelines, generating high-level navigation outputs that are fused at a centralized node. This approach minimizes real-time intersensor communication and computational complexity, making it well-suited for stable environments or systems with limited resources. However, its reliance on independent sensor outputs makes it less effective in dynamic or complex scenarios, where changes in sensor performance can significantly impact the overall accuracy.

2. Tightly coupled integration mode

In contrast, TC integration achieves a deep fusion of sensor data with real-time information exchange, significantly enhancing the positioning accuracy and stability of the system in complex environments and when the performance of individual sensors is limited [50]. By tightly fusing data from multiple sensors, TC approaches provide a more integrated and robust solution, ensuring reliable performance even in dynamic and challenging scenarios. This high level of integration comes at the cost of increased computational demands and complexity, making TC methods ideal for applications where precision and stability are more critical than simplicity, such as in complex environments. In GNSS/INS systems, the INS not only performs self-positioning but also provides real-time motion state information that assists in the tracking of GNSS signals and error correction, ensuring positioning continuity and accuracy when satellite signals are weak [50,78]. Studies such as [78,79] focus on the research of TC GNSS and INS technologies based on MEMS technology. The postprocessing kinematic/INS tightly coupled mode provided by GINav demonstrates significant advantages in positioning accuracy and fixed solution rate, particularly in environments with severe signal blockage, where it can maintain continuous and accurate trajectory tracking [80].

In an INS/visual system, both visual observation constraints and the IMU kinematic model constraints are integrated into a nonlinear optimization problem for joint state estimation. The TC strategy fully exploits the intrinsic correlation and redundant information between the two sensors, enabling more precise and comprehensive data fusion and error correction. This approach is particularly effective in scenarios with complex and dynamic environments, significant variations in lighting conditions, or when IMU errors accumulate over prolonged operation, leading to increased drift [81].

In more advanced combinations, such as GNSS/INS/visual systems, the data association and constraints between the sensors are rigorously modeled and processed. By leveraging the complementarity of GNSS, IMU, and visual, the TC integration method directly fuses raw GNSS carrier phase and pseudo range measurements, IMU data, and visual features at the observation level using the extended Kalman filter (EKF). This approach effectively utilizes the multisensor information and rejects potential outlier measurements [82–84].

In a TC GNSS/INS/visual/LiDAR system, GNSS signals provide the absolute positioning reference and its data are tightly integrated with the outputs of the INS gyroscopes and accelerometers. At the same time, the data from the visual sensor and LiDAR are deeply integrated into this optimization process. These two types of sensors provide rich spatial geometric constraints and other forms of observational information, which, particularly in complex environments such as urban canyons, underground parking garages, or areas with weak GNSS signals, can complement and enhance the navigation system's state estimation. This significantly improves the accuracy of motion trajectory tracking and the reliability of attitude perception [85]. However, this also demands more computational resources and increases algorithmic complexity. To address this, Li et al. adopted a semitight coupling strategy that combines GNSS PPP with S-VINS. This approach implements a bidirectional interaction mechanism, allowing the high-precision local positioning data generated by the S-VINS system to be exchanged and shared in real-time with the PPP system, creating complementary advantages. This not only maintains high stability but also enables centimeter-level or even subcentimeter-level positioning accuracy [86].

3. Deeply coupled integration mode

The concept of DL was first introduced in 1996 by Gustafson at the Draper Laboratory in the United States. He was the first to use the term "deeply coupled" to describe a combined navigation method that employs an extended code-tracking loop [87]. However, as research on deeply and ultratightly coupled has progressed, differing interpretations and understandings of these two concepts have emerged within both academic and industry circles. Two main perspectives currently prevail. One perspective holds that ultratightly and deeply coupled are essentially the same, both representing advanced techniques in which deeply integrated INS information aids the tracking loops of GNSS receivers. Another view, however, makes a more nuanced distinction, arguing that ultratightly coupled is a subset of deeply coupled. Specifically, deeply coupled systems can be divided into scalar and vector deeply coupled based on the type of receiver tracking loops used, with ultratightly coupled corresponding to vector deeply coupled. In this mode, the INS not only provides attitude and velocity information but also directly participates in the vector tracking loop of the GNSS receiver, achieving a higher level of fusion. Both viewpoints have their proponents but a consensus has yet to be reached within the industry. Researchers such as Liu [88] adopted the first perspective on deeply coupled GNSS and INS, innovatively introducing real-time carrier motion parameters provided by the IMU into the tracking loop assistance process. This approach successfully addressed the challenge of decreased GNSS carrier phase measurement accuracy in extreme environments. Wu et al. further analyzed and improved the prefiltering technology within deep integration structures and subsequently proposed an adaptive GNSS/INS deep integration algorithm featuring hybrid prefiltering processing [89].

In addition, the GNSS/IMU ultratight integration navigation system generally performs well in high-dynamic and harsh environments as it leverages the complementary strengths of both GNSS and IMU sensors, providing robust performance under challenging conditions. However, in situations where GNSS signals are severely degraded or obstructed, the system's performance may be affected, despite the assistance of IMU. To address this, researchers proposed a visually assisted GNSS/IMU ultratight integration navigation method. This method incorporates the relative positioning information provided by visual sensors into the conventional ultratight integration system, thereby enhancing the system's robustness and accuracy in situations where GNSS signals are interfered with or occluded [90,91].

In a word, DC approaches tightly integrate raw sensor measurements at the signal-tracking level within unified processing loops. This enables precise error correction and robust performance in dynamic and high-stress scenarios. However, the increased level of integration comes with greater computational demands and complexity, making DC approaches most suitable for applications requiring ultrahigh precision and stability, such as such as the navigation in urban canyons or UAVs operating under dense foliage, where sensor degradation and multisource interference are prevalent.

In summary, whether it be LC, TC, DL, or ultratightly coupled, each represents a different level of information fusion strategy within integrated navigation systems to enhance positioning accuracy and stability. The choice and application of these fusion methods must be based on the requirements of the specific application scenario, environmental conditions, and a balance between system complexity and cost. With continuous advancements in sensor technology, improved computational power, and the integration of artificial intelligence, especially the development and application of advanced fusion methods like deeply and ultratightly coupled, integrated navigation systems are poised to become more precise and intelligent. This evolution will better meet the complex environmental and high-precision positioning demands of cutting-edge fields such as autonomous driving and unmanned aerial vehicles.

4.2.2. State Estimation Methods in Integrated Navigation

State estimation lies at the core of integrated navigation technology, directly affecting the accuracy and robustness of navigation systems in complex dynamic environments. Figure 12 clearly illustrates the three key technologies in integrated navigation state estimation: filtering techniques, graph optimization methods, and deep learning. As shown in

Figure 12, prior to 2018, Kalman filtering [30,92–95] and its derivative algorithms [96–99] dominated state estimation and data fusion in navigation. These algorithms, with their rigorous mathematical frameworks and efficient online processing capabilities, provided a solid foundation for high-precision positioning [96,98,99]. However, as environmental complexity increased, data dimensions expanded, and computational capabilities advanced, the limitations of these classical methods in handling nonlinearity, high noise, and model uncertainties have gradually become apparent. Since 2013, researchers have begun exploring optimization-based approaches for integrated navigation to overcome the limitations of traditional filtering algorithms and enhance system robustness and accuracy [100–104]. Around 2020, optimization methods, specifically graph optimization techniques, began to gain significant attention in integrated navigation, further advancing the field's focus on refined optimization approaches. By 2022, deep learning technologies, with their powerful capabilities in feature extraction, nonlinear modeling, and adaptive learning, effectively addressed challenges faced by traditional methods. In particular, deep learning contributed significantly to the intelligence and accuracy of integrated navigation systems in complex data processing, environmental adaptability, and noise suppression [13]. The deep integration of deep learning with integrated navigation technology opens up vast prospects for future research, heralding a new era of highly intelligent, high-precision navigation.

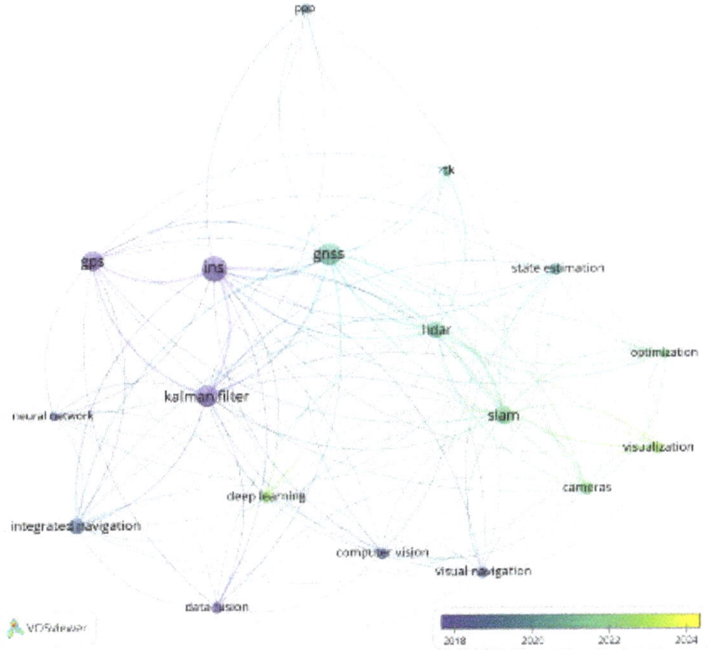

Figure 12. Timeline of research hotspots in integrated navigation algorithms.

1. Filtering techniques in integrated navigation.

The origins of the Kalman filter can be traced back to 1960, when R. E. Kalman introduced this groundbreaking algorithm [105]. Over time, it has become integral in GNSS/INS integrated systems, where it integrates positioning and navigation data to provide accurate estimates of position, velocity, and attitude through prediction and update steps while enhancing the system's robustness in the absence of GPS signals [106]. To further improve the performance of GNSS/INS, scientists have proposed EKF, particle filtering, and UKF techniques [107]. The EKF algorithm performs a first-order Taylor expansion for nonlinear systems and is suitable for weakly nonlinear systems but it is less effective when dealing with strongly nonlinear systems. While particle filtering and UKF can handle strongly nonlinear systems more effectively, they come with higher algorithmic complexity and computational resource demands [108,109]. To address these issues, the paper [110] presents an improved robust strong tracking unscented Kalman filter (RSTUKF), which specifically enhances the filtering capability of the system in the presence of dynamic model errors, reducing the adverse effects of prior information. Research in 2024 has further expanded the application of these techniques in integrated navigation. Xu et al. combined a GRU neural network with adaptive Kalman filtering (AKF) to dynamically adjust filter parameters during GNSS signal outages, significantly improving the system's positioning accuracy [111]. Furthermore, Zhang et al. proposed an improved INS-level fusion algorithm that uses a shared covariance matrix to optimize the fusion of IMU arrays and GNSS data, effectively enhancing the consistency of the filtering results and computational efficiency [112]. These studies not only address the limitations of traditional filtering methods in strong nonlinear scenarios but also further enhance positioning accuracy and system efficiency through novel algorithm designs, driving the advancement of integrated navigation technology.

In a VINS system, the filtering algorithm integrates the continuous, high-frequency data from the IMU, which suffers from cumulative errors, with the intermittent, low-frequency but highly accurate environmental feature information obtained from the visual sensor in order to precisely estimate the system's state parameters. However, filtering-based systems are limited by the constraints of state variable selection, the complexity of constructing state and observation equations, and insufficient capability to handle nonlinear and noisy data [113]. To overcome these limitations, researchers have integrated IMU biases and camera parameters within the multistate constraint Kalman filter (MSCKF) framework, thereby improving data fusion accuracy [114]. Hesch et al. improved the filter's performance in handling system rotation dynamics by identifying and addressing unobservable modes in nonlinear systems [115]. Meanwhile, R-VIO [116] and MIMC-VINS [117] enhance the accuracy and robustness of state estimation in complex dynamic environments by jointly processing visual and IMU data.

In the development of GNSS/INS/visual integrated navigation technology [101], the MSF-EKF enhances the flexibility of sensor configuration through modular design but it still has limitations when dealing with highly nonlinear dynamics [118]. The article [119] proposes an adaptive federated filter that fuses GNSS, INS, and visual odometry (VO) data, effectively addressing nonlinear issues in high-dynamic environments. By adaptively adjusting the weights, this method improves the system's accuracy and robustness in complex settings. However, when the dynamics of the environment become more extreme, traditional filtering methods often encounter convergence issues. Researchers further proposed a GNSS/INS/visual integration method based on the invariant extended Kalman filter (IEKF). This method introduces system dynamics invariants, effectively reducing linearization errors and significantly improving convergence speed and system robustness in high-dynamic environments, demonstrating exceptional performance in complex scenarios [120]. To further improve positioning accuracy and system stability, Liao et al. improved positioning accuracy in GNSS signal-constrained environments by combining EKF with VIO [121]. Gu et al. further improved data fusion accuracy and robustness by introducing cascade Kalman filtering and dynamic object removal algorithms [122]. Additionally, with the increasing demand for high-precision and high-reliability GNSS/INS/visual integrated navigation technologies, the InGVIO algorithm was proposed. By utilizing invariant filtering, it ensured filtering performance under specific motion patterns, thereby enhancing navigation accuracy in complex scenarios [82,123]. In the realm of integrity monitoring, researchers have proposed a solution based on the error state EKF model. This approach not only estimates the system's fundamental state but also explicitly models the system errors, thereby enhancing fault detection and system response capabilities. By incorporating the error state model, this solution significantly improves the robustness and reliability of the system in complex and dynamic environments.

Furthermore, in the integration of GNSS/INS/visual/LiDAR systems, GNSS data is tightly coupled with the outputs of the INS gyroscope and accelerometer. Filtering techniques like EKF or UKF [124–127] employ real-time error estimation and compensation mechanisms to synchronize sensor data processing. When GNSS signals are weak or lost, INS can provide inertial navigation data to compensate. Additionally, the fusion of visual and LiDAR data further enhances the system's robustness and accuracy in complex environments. Researchers optimize the processing of point cloud data from both visual and LiDAR sensors, using the visual sensor for feature extraction and recognition in static environments, while LiDAR provides high-precision three-dimensional spatial information. Through joint filtering, these sensors complement one another, reducing individual error accumulation and thus improving the overall system accuracy and reliability. Particularly in dynamic environments, the high resolution of LiDAR and the rich texture information from visual sensors effectively mitigate positioning errors caused by GNSS signal loss [85].

2. Graph optimization methods in integrated navigation

The graph optimization algorithm estimates state variables using probabilistic graphical models, making it particularly well-suited for handling asynchronous data fusion. By constructing a highly flexible factor graph framework, it enables true "plug-and-play" integration of sensors [128–132]. In GNSS/INS, sliding window and factor graph models are used to reduce computational burdens, while INS redundancy is leveraged to detect and exclude anomalous GNSS data, enhancing the system's robustness and accuracy [133]. To solve the problem of IMU installation angle estimation, a hybrid factor graph method based on nonlinear optimization has been proposed. By integrating a loosely coupled GNSS/INS algorithm with an optimized objective function, this method effectively reduces errors in the IMU installation angle, significantly improving three-dimensional positioning accuracy [134]. Further research has also led to the optimization of the inertial preintegration model. By introducing a high-precision IMU factor that accounts for Earth's rotation, combined with factor graph optimization, the performance of the inertial navigation system in attitude estimation has been significantly enhanced [135]. Moreover, in terms of short-term relative accuracy, factor graph optimization has shown improvements in both three-dimensional position and velocity accuracy compared to traditional EKF methods. This approach is particularly well-suited for applications that require short-term high-precision navigation, such as railway inspection and mobile mapping [136].

In VINS, graph optimization and factor graph optimization are also widely applied. By constructing nonlinear optimization problems, these methods effectively address complex nonlinear relationships and cumulative errors, achieving globally optimal state estimation. The incremental smoothing and mapping (iSAM) algorithm [137] used factor graphs for real-time smoothing and mapping, efficiently processing large-scale data and enhancing the global consistency and accuracy of the VINS system. Building on this, Indelman et al. proposed an incremental smoothing framework that allows dynamic updates of sensor data, improving system flexibility and adaptability to complex environments [132]. RD-VIO further utilizes IMU-PARSAC for robust keypoint detection and matching and addresses the pure rotation problem through delayed triangulation, which effectively enhances navigation performance in dynamic environments [138]. Additionally, DynaVIO employs a dynamic probabilistic propagation model to identify and reject dynamic features, thereby improving the robustness of position estimation [139]. In loop closure detection and global map consistency maintenance, graph optimization also plays a key role. For instance, loop closure processing in the MonoSLAM system [66] inspired subsequent optimization strategies in systems like the ORB-SLAM series [28,31,35], where precise graph optimization enables trajectory optimization and global relocalization, enhancing the long-term stability and accuracy of VINS. Additionally, algorithms such as VINS-Mono [24] and VINS-Fusion combine IMU preintegration techniques with nonlinear optimization of visual data, significantly improving system performance in dynamic and complex environments. The development of these technologies has made VINS systems more accurate and stable in handling large-scale data and navigating complex environments.

In GNSS/INS/visual systems, the GOMSF method proposed by Mascaro et al. enhanced the flexibility and robustness of multisensor data fusion through a sliding window pose graph optimization model [140]. The IC-GVINS algorithm [141] further strengthened the application of INS-based real-time navigation systems in complex environments by efficiently integrating multisensor data through the combination of factor graph optimization and sliding window techniques. In the same year, the open-source GICI-LIB library [142] and the Sky-GVINS algorithm [143] improved the performance of navigation systems in dynamic and signal-constrained environments by considering GNSS errors and leveraging visual data. Figure 13 illustrates a typical factor graph optimization framework for GNSS/IMU/visual/LiDAR integration. In this integration, optimization techniques demonstrate unique advantages in addressing complex nonlinear problems and accumulated errors. The NDT SLAM algorithm proposed by Ravi et al. combined graph optimization with deep learning, significantly improving trajectory accuracy and automation levels [144]. Zhou et al. further applied these techniques to UAV platforms, achieving centimeter-level mapping accuracy through precise point cloud matching and SFM strategies [145]. The GVIL algorithm from Wuhan University enhanced navigation performance in complex environments by using a tightly coupled graph optimization approach, overcoming challenges in accuracy and continuity under GNSS-limited conditions, demonstrating the potential for seamless navigation [146].

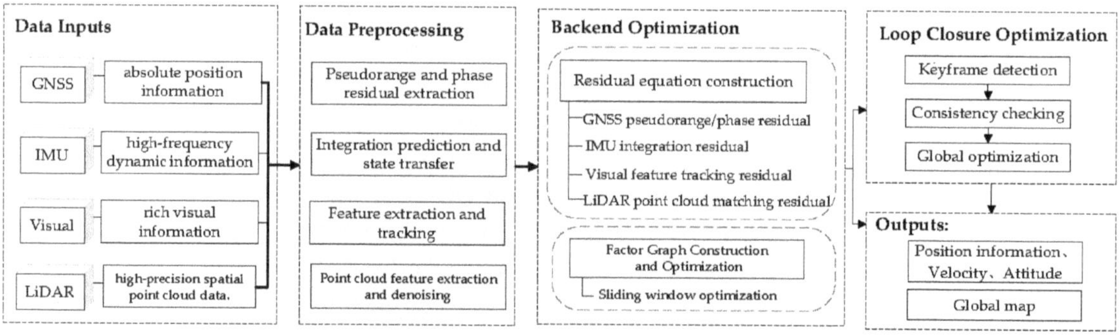

Figure 13. GNSS/IMU/visual/LiDAR multisensor fusion factor graph optimization framework.

3. Deep learning techniques in integrated navigation

The nonlinear modeling capability of neural networks provides new solutions for GNSS/INS integrated navigation. In navigation systems, traditional linear filtering methods (such as EKF) have limitations when handling strong nonlinear dynamic relationships. In contrast, neural networks, through deep analysis and adaptive adjustment of multisource data, are capable of dynamically capturing the nonlinear relationship between GNSS and INS data, demonstrating significant potential in feature modeling and error compensation in complex scenarios. Liu et al. explored the application of LSTM in GNSS/INS integration, compensating for INS errors using pseudo-GNSS position information, which significantly improved navigation performance during signal interruptions [147]. Wu et al. proposed a deep-learning-based adaptive error compensation method, dynamically correcting GNSS accumulated errors caused by signal interruptions using INS data. Compared to traditional extended Kalman filtering, the navigation accuracy improved by 77.7% [148]. Additionally, researchers have combined deep learning with extended Kalman filtering and adaptive Kalman filtering (AKF), dynamically adjusting filtering parameters during GNSS signal loss, significantly improving system positioning accuracy [149,150]. In recent years, neural networks have been further applied in complex environments such as urban canyons and multipath interference, integrating technologies like CNN and transformer, driving the widespread application of GNSS/INS integrated navigation systems in autonomous driving, UAV navigation, and robotic positioning.

In the development of VINS, deep learning has gradually become an important tool for enhancing system autonomy, resilience to interference, and navigation accuracy. Long short-term memory (LSTM) networks are used to learn the temporal sequence characteristics of IMU data, improving the state transition model, while CNN enhances the observation model's accuracy by extracting multilevel features from visual images. These deep learning modules replace certain traditional filters, reducing errors introduced by simplifying mathematical models [151]. Similarly, the DynaNet proposed by Chen et al. utilizes LSTM to estimate the state transition matrix, combined with CNN to extract high-level features, thereby constructing KF equations to improve the accuracy of state estimation [152]. Additionally, DeepVIO combines self-supervised learning with 3D geometric constraints, utilizing 2D optical flow features and IMU data for trajectory estimation without relying on external calibration data. This approach maintains high robustness in the presence of calibration errors and data loss [153]. On the other hand, VIFT employs a transformer-based deep learning model, utilizing an attention mechanism for pose estimation, optimizing the temporal dependencies of latent feature vectors, and performing rotation learning on the SE (3) manifold while addressing data imbalance. This approach is particularly well-suited for monocular visual–inertial navigation systems in autonomous driving and robotics [154]. Deep learning is also applied to the real-time evaluation of the confidence level of observational data. When encountering outlier data or sensor failures, the filter can intelligently perform data selection and weight assignment, mitigating the impact of erroneous data [155].

Deep learning has also demonstrated immense potential in GNSS/INS/visual integrated navigation, capable of handling signal attenuation in complex environments and dynamically adjusting sensor weights. By combining CNN for visual feature extraction and RNN/LSTM for capturing IMU temporal dependencies, this approach is particularly suited for scenarios with weak or interrupted GNSS signals. Although research is still in its early stages, systems like VD-VIO [156] have already shown the promising prospects of deep learning in these areas. In further research, a deep-learning-assisted fusion technique for inertial, visual, and LiDAR data is introduced, where CNN is employed to extract spatial features from vision and LiDAR while GRU captures the temporal dependencies of IMU data. The method dynamically adjusts the fusion weights of the different data modalities and detects and rejects anomalous sensor data to ensure the reliability of the multimodal fusion process. With advancements in algorithms and hardware, the integration of deep learning and integrated navigation technologies is expected to address challenges such as positioning in extreme environments, rapid adaptation to new environments, and navigation in weak signal conditions, thus driving further development in the field of navigation. This trend signals the arrival of a new era of deep intelligence and high-precision navigation, opening up vast prospects for future research.

In summary, the three key technologies for state estimation in integrated navigation each have their own advantages and applicable scenarios, so a comprehensive consideration of various factors is required when choosing an algorithm. For simple and stable environments with limited computational resources, filtering techniques are more suitable. In cases where data processing, robustness, and fault detection are emphasized and hardware resources are sufficient, graph optimization methods are superior. In complex urban environments, particularly when GNSS signals are frequently interrupted and data accumulation occurs, deep learning techniques demonstrate clear advantages. Table 6 provides a detailed summary of the data fusion algorithms for the integration of GNSS, INS, visual, and LiDAR.

Table 6. Summary of four typical integrated navigation modes and their algorithms.

Sensor Type	Algorithm Name	Fusion Type	Algorithm Type	Features
GNSS/INS	KF-GINS	LC	EKF	Adopts optimized Bayesian methods with high robustness; effectively processes noise, outliers, and uncertainties
	OB_GINS	LC	Graph Optimization	Optimized Bayesian approach, enhances robustness against noise, outliers, and uncertainties.
	GINav	LC/TC	EKF	Applications for real-time and postprocessing; flexible sensor configurations and an extensible framework for advanced filtering and other sensor integrations.
	Ignav	LC/TC	Filtering	Modular design, adaptable to various application scenarios; high flexibility.
INS/visual	ROVIO	LC	IEKF	High real-time performance and robustness; suitable for fast-moving platforms.
	MSCKF	TC	KF	Suitable for embedded platforms; lightweight real-time navigation, suppresses motion blur and lighting changes.
	MSCEqF	TC	Equivariant Filtering	Integrates multiple state variables with self-calibration capabilities; applies group theory for equivariant design, ensuring improved linearization and internal consistency.
	R-VIO	TC	EKF	Uses a mobile local coordinate system for state estimation; operates from any initial posture without requiring alignment with the global gravity direction.
	MIMC-VINS	TC	Multistate Constrained Kalman Filter	Fuses data from multiple uncalibrated cameras and IMUs; ensures seamless 3D motion tracking; incorporates on-manifold state interpolation and online calibration for spatiotemporal and intrinsic parameters.
	VINS-Mono	TC	Graph Optimization	Monocular vision and inertial fusion; preintegration and partial marginalization techniques for large-scale graph optimization.
	OKVIS	TC	Graph Optimization	Keyframe-based global pose estimation; handles scale uncertainty, suitable for complex dynamic environments.
	StructVIO	TC	Graph Optimization	Structured line features with iterative optimization graph optimization; improves accuracy in weak and repetitive texture environments.

Table 6. *Cont.*

Sensor Type	Algorithm Name	Fusion Type	Algorithm Type	Features
	ORB-SLAM3	TC	Graph Optimization	Multisensor support (monocular, stereo, RGB-D); visual + IMU tightly coupled; multimap system; excellent positioning accuracy and robustness.
	RD-VIO	TC	Graph Optimization	Utilizes IMU-PARSAC for robust keypoint detection and matching; employs deferred triangulation to handle pure rotation; enhances performance in dynamic environments.
INS/visual	DFF-VIO	TC	Graph Optimization	Utilizes dynamic feature matching, handles dynamic environments, improves tracking accuracy and robustness, especially for moving objects.
	DynaVIO	TC	Graph Optimization	Identifies and rejects dynamic features; uses dynamic probability propagation for robust pose estimation.
	DeepVIO	TC	Deep Learning (LSTM + CNN + Fusion Network)	Combines 2D optical flow and IMU data; self-supervised learning for trajectory estimation; robust to calibration errors and missing data.
	VIFT	TC	Deep Learning (Transformer-Based)	Uses attention mechanisms for pose estimation; refines latent feature vectors temporally; addresses data imbalance and rotation learning in SE(3); suitable for monocular VIO systems in autonomous driving and robotics.
GNSS/INS/ visual	IC-GVINS	GNSS LC	Factor Graph Optimization	Adapted to complex environments and sensor failures.
	InGVIO	GNSS TC	IEKF	High precision and efficient computation with keyframe marginalization; robust performance.
	GICI-LIB	GNSS LC/TC	Factor Graph Optimization	The first open-source integrated navigation library; multifrequency processing and optimized RTK, PPP algorithms for high-precision fusion navigation.
	GVINS	GNSS TC	Graph Optimization	Accurate and robust 3D navigation and positioning, suitable for high-precision and high-robustness requirements.
	Sky-GVINS	GNSS TC	Graph Optimization	Optimized for UAVs; high-precision, low-latency 3D positioning and attitude estimation.
GNSS/INS/ visual/LiDAR	GVIL	TC	Graph Optimization	Integrates multisensor information for high-precision, comprehensive 3D perception and positioning, suitable for complex environments and autonomous driving.

Future research will focus on several key directions. First, improving sensor quality remains fundamental, particularly in enhancing the stability, temperature adaptability, and environmental robustness of inertial sensors (e.g., MEMS inertial sensors). Recent studies have shown that employing noise compensation and temperature compensation techniques can significantly improve the accuracy of MEMS sensors [157–159]. Meanwhile, the adaptability of visual sensors in complex lighting conditions and high-speed motion scenarios has been progressively improved through deep learning algorithms, such as CNN-informer. Second, the trend of hardware integration and miniaturization is driving navigation devices toward compact, low-power, and cost-effective solutions [160,161]. For example, the widespread adoption of wearable devices and embedded navigation modules has opened up new application scenarios for navigation technologies [160]. Third, the optimization of information fusion algorithms remains central to technological advancement. Future efforts will focus on improving the real-time performance of algorithms and enhancing the system's adaptive adjustment capabilities, thereby increasing the reliability and practicality of navigation systems in dynamic and uncertain environments. The development of hybrid navigation strategies to achieve deep integration and complementarity of the three key technologies will be a priority. For example, flexibly combining filtering and optimization techniques allows GNSS and INS data to be processed efficiently through filtering while incorporating graph optimization with visual information ensures optimal performance across diverse scenarios. Deep learning enhances robustness in real-time parameter calibration, environmental perception, and system adaptability. Its integration with traditional filtering methods can further improve system accuracy and adaptability. The combination of factor graph optimization and deep learning enhances system adaptability through real-time feature extraction and dynamic weight adjustment. Finally, the emergence of vision–language navigation has injected innovative vitality into navigation systems. Future research could explore the integration of vision–language navigation with multisensor fusion technologies, fully harnessing the potential of visual and language information to further enhance the decision-making efficiency and accuracy of navigation systems in dynamic environments. This trend heralds a new era of deeply intelligent navigation, offering greater potential for the application of navigation technologies in complex environments.

5. Conclusions

With the rapid development of integrated navigation technologies, significant progress has been made in the fusion of multisource sensors. This paper provides a literature review of research on sensor combinations, including GNSS, INS, visual, and LiDAR, spanning from 2000 to 2024. It outlines the current state, research gaps, and future development trends of mainstream integrated navigation systems such as GNSS/INS, INS/visual, GNSS/INS/visual, and GNSS/INS/visual/LiDAR. Despite the remarkable performance of these technologies in improving navigation accuracy and environmental adaptability, several challenges remain in practical applications. For instance, in complex urban environments, the combination of INS and visual proves to be more advantageous, whereas in open environments, the integration of GNSS and INS performs better. While LiDAR excels in high-precision modeling, its data collection reliability under adverse weather conditions still requires further improvement. In the future, optimizing multisensor fusion algorithms, enhancing real-time data processing capabilities, and developing adaptive adjustment strategies will be key to improving system performance. Additionally, the integration of advanced technologies such as deep learning and neural networks will further enhance the intelligence and automation levels of integrated navigation systems. The ideal future integrated navigation system should be capable of flexibly leveraging the advantages of various sensors based on environmental conditions and seamlessly integrating them through efficient data fusion algorithms. This will ensure that the navigation system maintains high accuracy and reliability even in the event of a sensor failure or environmental changes. This developmental trend will drive the advancement of integrated navigation technology toward higher precision and broader environmental adaptability. To achieve this goal, researchers must focus on overcoming key challenges such as enhancing GNSS resistance to interference, improving the INS error model and correction algorithms to suppress drift, optimizing the robustness of visual sensors under complex conditions, and addressing the performance limitations of LiDAR in adverse weather. Additionally, the development of new LiDAR technologies with better penetration capabilities or the design of more efficient environmental-adaptive data processing algorithms will be essential areas of focus for future research.

Looking ahead, the development of integrated navigation systems will increasingly focus on multisensor fusion and intelligent optimization. The introduction of emerging sensor technologies, such as quantum navigation and bionavigation, along with the application of algorithms like deep learning, will drive the widespread adoption of this technology in fields such as autonomous driving, aerospace, and robotics. These advancements will ensure precise navigation in extremely complex environments.

Author Contributions: Conceptualization, J.W.; methodology, J.W.; validation, J.W.; data curation, J.W.; writing and original draft preparation, J.W.; writing, reviewing, and editing, M.S. and Y.Y.; visualization, J.W.; supervision, Y.Y.; funding acquisition, M.S. All authors have read and agreed to the published version of the manuscript.

Funding: This work was supported by the Hubei Strategic Science and Technology Talent Program (No. KJCXRC202400130), the Natural Science Foundation of Hubei Province of China (No. 2023AFB948), the Project of the State Key Laboratory of Geodesy and Earth's Dynamics (No. S22L6201), and the National Key Research Program (No. 2022YFB3903903).

Conflicts of Interest: The authors declare no conflicts of interest.

References

1. Adeel, M.; Gong, Z.; Liu, P.L.; Wang, Y.Z.; Chen, X.; Inst, N. Research and Performance Analysis of Tightly Coupled Vision, INS and GNSS System for Land Vehicle Applications. In Proceedings of the 30th International Technical Meeting of The Satellite-Division-of-the-Institute-of-Navigation (ION GNSS+), Portland, OR, USA, 25–29 September 2017; pp. 3321–3330.
2. Chiang, K.W.; Chang, H.W.; Li, Y.H.; Tsai, G.J.; Tseng, C.L.; Tien, Y.C.; Hsu, P.C. Assessment for INS/GNSS/Odometer/Barometer Integration in Loosely-Coupled and Tightly-Coupled Scheme in a GNSS-Degraded Environment. *IEEE Sens. J.* **2020**, *20*, 3057–3069.
3. Geng, X.S.; Guo, Y.; Tang, K.H.; Wu, W.Q.; Ren, Y.C. Research on Covert Directional Spoofing Method for INS/GNSS Loosely Integrated Navigation. *IEEE Trans. Veh. Technol.* **2023**, *72*, 5654–5663.
4. Cong, L.; Yue, S.; Qin, H.L.; Li, B.; Yao, J.T. Implementation of a MEMS-Based GNSS/INS Integrated Scheme Using Supported Vector Machine for Land Vehicle Navigation. *IEEE Sens. J.* **2020**, *20*, 14423–14435.
5. Shang, X.Y.; Sun, F.P.; Liu, B.D.; Zhang, L.D.; Cui, J.Y. GNSS Spoofing Mitigation With a Multicorrelator Estimator in the Tightly Coupled INS/GNSS Integration. *IEEE Trans. Instrum. Meas.* **2023**, *72*, 12.
6. Wang, J.; Wang, S.; Zou, D.; Chen, H.; Zhong, R.; Li, H.; Zhou, W.; Yan, K. Social Network and Bibliometric Analysis of Unmanned Aerial Vehicle Remote Sensing Applications from 2010 to 2021. *Remote Sens.* **2021**, *13*, 2912. [CrossRef]
7. Huang, Z.; Wu, X.; Wang, H.; Hwang, C.; He, X. Monitoring Inland Water Quantity Variations: A Comprehensive Analysis of Multi-Source Satellite Observation Technology Applications. *Remote Sens.* **2023**, *15*, 3945. [CrossRef]
8. Van Eck, N.; Waltman, L. Software survey: VOSviewer, a computer program for bibliometric mapping. *Scientometrics* **2010**, *84*, 523–538. [PubMed]
9. Wei, J.; Wang, F.; Lindell, M.K. The evolution of stakeholders' perceptions of disaster: A model of information flow. *J. Assoc. Inf. Sci. Technol.* **2016**, *67*, 441–453.
10. Chen, C.; Dubin, R.; Kim, M.C. Emerging trends and new developments in regenerative medicine: A scientometric update (2000–2014). *Expert Opin. Biol. Ther.* **2014**, *14*, 1295–1317.
11. Schmoch, U. Mean values of skewed distributions in the bibliometric assessment of research units. *Scientometrics* **2020**, *125*, 925–935.
12. Chen, C.M. Eugene Garfield's scholarly impact: A scientometric review. *Scientometrics* **2018**, *114*, 489–516. [CrossRef]
13. Yang, L.; Li, Y.; Wu, Y.L.; Rizos, C. An enhanced MEMS-INS/GNSS integrated system with fault detection and exclusion capability for land vehicle navigation in urban areas. *GPS Solut.* **2014**, *18*, 593–603. [CrossRef]
14. Bhatti, U.I.; Ochieng, W.Y.; Feng, S.J. Integrity of an integrated GPS/INS system in the presence of slowly growing errors. Part I: A critical review. *GPS Solut.* **2007**, *11*, 173–181. [CrossRef]
15. Luo, Q.; Cao, Y.R.; Liu, J.J.; Benslimane, A. Localization and Navigation in Autonomous Driving: Threats and Countermeasures. *IEEE Wirel. Commun.* **2019**, *26*, 38–45. [CrossRef]
16. Hollar, S.; Brain, M.; Nayak, A.A.; Stevens, A.; Patil, N.; Mittal, H.; Smith, W.J. A New Low Cost, Efficient, Self-Driving Personal Rapid Transit System. In Proceedings of the 28th IEEE Intelligent Vehicles Symposium (IV), Redondo Beach, CA, USA, 11–14 June 2017; pp. 412–417.
17. Li, S.; Cui, P.Y.; Cui, H.T. Autonomous navigation and guidance for landing on asteroids. *Aerosp. Sci. Technol.* **2006**, *10*, 239–247. [CrossRef]
18. Sabatini, R.; Moore, T.; Ramasamy, S. Global navigation satellite systems performance analysis and augmentation strategies in aviation. *Prog. Aeosp. Sci.* **2017**, *95*, 45–98. [CrossRef]

19. Walter, T.; Enge, P.; Blanch, J.; Pervan, B. Worldwide Vertical Guidance of Aircraft Based on Modernized GPS and New Integrity Augmentations. *Proc. IEEE* **2008**, *96*, 1918–1935. [CrossRef]
20. Chang, L.; Niu, X.J.; Liu, T.Y.; Tang, J.; Qian, C. GNSS/INS/LiDAR-SLAM Integrated Navigation System Based on Graph Optimization. *Remote Sens.* **2019**, *11*, 1009. [CrossRef]
21. Yang, Z.; Yu, X.; Dedman, S.; Rosso, M.; Zhu, J.; Yang, J.; Xia, Y.; Tian, Y.; Zhang, G.; Wang, J. UAV remote sensing applications in marine monitoring: Knowledge visualization and review. *Sci. Total Environ.* **2022**, *838*, 155939.
22. Noureldin, A.; Karamat, T.B.; Eberts, M.D.; El-Shafie, A. Performance Enhancement of MEMS-Based INS/GPS Integration for Low-Cost Navigation Applications. *IEEE Trans. Veh. Technol.* **2009**, *58*, 1077–1096. [CrossRef]
23. Skog, I.; Handel, P. In-Car Positioning and Navigation Technologies—A Survey. *IEEE Trans. Intell. Transp. Syst.* **2009**, *10*, 4–21. [CrossRef]
24. Qin, T.; Li, P.; Shen, S. VINS-Mono: A Robust and Versatile Monocular Visual-Inertial State Estimator. *IEEE Trans. Robot.* **2018**, *34*, 1004–1020. [CrossRef]
25. Forster, C.; Carlone, L.; Dellaert, F.; Scaramuzza, D. On-manifold preintegration for real-time visual--inertial odometry. *IEEE Trans. Robot.* **2017**, *33*, 1–21.
26. Xu, W.; Cai, Y.X.; He, D.J.; Lin, J.R.; Zhang, F. FAST-LIO2: Fast Direct LiDAD-Inertial Odometry. *IEEE Trans. Robot.* **2022**, *38*, 2053–2073. [CrossRef]
27. Geiger, A.; Lenz, P.; Stiller, C.; Urtasun, R. Vision meets robotics: The kitti dataset. *Int. J. Robot. Res.* **2013**, *32*, 1231–1237. [CrossRef]
28. Campos, C.; Elvira, R.; Rodriguez, J.J.G.; Montiel, J.M.M.; Tardos, J.D. ORB-SLAM3: An Accurate Open-Source Library for Visual, Visual–Inertial, and Multimap SLAM. *IEEE Trans. Robot.* **2021**, *37*, 1874–1890. [CrossRef]
29. Paull, L.; Saeedi, S.; Seto, M.; Li, H. AUV Navigation and Localization: A Review. *IEEE J. Ocean. Eng.* **2014**, *39*, 131–149. [CrossRef]
30. Li, M.; Mourikis, A.I. High-precision, consistent EKF-based visual-inertial odometry. *Int. J. Robot. Res.* **2013**, *32*, 690–711.
31. Mur-Artal, R.; Tardós, J.D. Visual-Inertial Monocular SLAM With Map Reuse. *IEEE Robot. Autom. Lett.* **2017**, *2*, 796–803. [CrossRef]
32. Chen, C. CiteSpace II: Detecting and visualizing emerging trends and transient patterns in scientific literature. *J. Am. Soc. Inf. Sci. Technol.* **2006**, *57*, 359–377.
33. Groves, P.D. *Principles of GNSS, Inertial, and Multisensor Integrated Navigation Systems*, 2nd ed.; Artech: Boston, MA, USA, 2013; pp. 1–776.
34. Leutenegger, S.; Lynen, S.; Bosse, M.; Siegwart, R.; Furgale, P. Keyframe-based visual–inertial odometry using nonlinear optimization. *Int. J. Robot. Res.* **2015**, *34*, 314–334. [CrossRef]
35. Mur-Artal, R.; Montiel, J.M.M.; Tardos, J.D. ORB-SLAM: A Versatile and Accurate Monocular SLAM System. *IEEE Trans. Robot.* **2015**, *31*, 1147–1163. [CrossRef]
36. Mur-Artal, R.; Tardós, J.D. ORB-SLAM2: An Open-Source SLAM System for Monocular, Stereo, and RGB-D Cameras. *IEEE Trans. Robot.* **2017**, *33*, 1255–1262. [CrossRef]
37. Jazwinski, A.H. *Stochastic Processes and Filtering Theory*; Courier Corporation: Chelmsford, MA, USA, 2007.
38. Sinpyo, H.; Man Hyung, L.; Ho-Hwan, C.; Sun-Hong, K.; Speyer, J.L. Observability of error States in GPS/INS integration. *IEEE Trans. Veh. Technol.* **2005**, *54*, 731–743. [CrossRef]
39. Groves, P.D. *Principles of GNSS, Inertial, and Multisensor Integrated Navigation Systems*; Artech: Boston, MA, USA, 2008; pp. 1–518.
40. Farrell, J. *Aided Navigation: GPS with High Rate Sensors*; McGraw-Hill, Inc.: New York, NY, USA, 2008.
41. Georgy, J.; Noureldin, A.; Korenberg, M.J.; Bayoumi, M.M. Low-Cost Three-Dimensional Navigation Solution for RISS/GPS Integration Using Mixture Particle Filter. *IEEE Trans. Veh. Technol.* **2010**, *59*, 599–615. [CrossRef]
42. Kaplan, E.D.; Hegarty, C. *Understanding GPS/GNSS: Principles and Applications*; Artech house: Boston, MA, USA, 2017.
43. Martinelli, A. Vision and IMU Data Fusion: Closed-Form Solutions for Attitude, Speed, Absolute Scale, and Bias Determination. *IEEE Trans. Robot.* **2012**, *28*, 44–60. [CrossRef]
44. Zhang, T.; Xu, X. A new method of seamless land navigation for GPS/INS integrated system. *Measurement* **2012**, *45*, 691–701. [CrossRef]
45. Chen, X.; Shen, C.; Zhang, W.-b.; Tomizuka, M.; Xu, Y.; Chiu, K. Novel hybrid of strong tracking Kalman filter and wavelet neural network for GPS/INS during GPS outages. *Measurement* **2013**, *46*, 3847–3854. [CrossRef]
46. Chowdhary, G.; Johnson, E.N.; Magree, D.; Wu, A.; Shein, A. GPS-denied indoor and outdoor monocular vision aided navigation and control of unmanned aircraft. *J. Field Robot.* **2013**, *30*, 415–438. [CrossRef]
47. Engel, J.; Schöps, T.; Cremers, D. LSD-SLAM: Large-Scale Direct Monocular SLAM. In Proceedings of the Computer Vision—ECCV 2014, Cham, Switzerland, 6–12 September 2014; pp. 834–849.
48. Forster, C.; Pizzoli, M.; Scaramuzza, D. SVO: Fast semi-direct monocular visual odometry. In Proceedings of the 2014 IEEE International Conference on Robotics and Automation (ICRA), Hong Kong, China, 31 May–7 June 2014; pp. 15–22.
49. Yang, Z.; Shen, S. Monocular Visual–Inertial State Estimation With Online Initialization and Camera–IMU Extrinsic Calibration. *IEEE Trans. Autom. Sci. Eng.* **2017**, *14*, 39–51. [CrossRef]

50. Falco, G.; Pini, M.; Marucco, G. Loose and Tight GNSS/INS Integrations: Comparison of Performance Assessed in Real Urban Scenarios. *Sensors* **2017**, *17*, 255. [CrossRef] [PubMed]
51. Usenko, V.; Engel, J.; Stückler, J.; Cremers, D. Direct visual-inertial odometry with stereo cameras. In Proceedings of the 2016 IEEE International Conference on Robotics and Automation (ICRA), Stockholm, Sweden, 16–21 May 2016; pp. 1885–1892.
52. Sun, K.; Mohta, K.; Pfrommer, B.; Watterson, M.; Liu, S.; Mulgaonkar, Y.; Taylor, C.J.; Kumar, V. Robust Stereo Visual Inertial Odometry for Fast Autonomous Flight. *IEEE Robot. Autom. Lett.* **2018**, *3*, 965–972. [CrossRef]
53. Ning, J.; Yao, Y.; Zhang, X. Review of the Development of Global Satellite Navigation System. *J. Navig. Position.* **2013**, *1*, 3–8. [CrossRef]
54. Liu, Z.; Zhou, Q.; Qin, Y.; El-Sheimy, N. Vision-Aided Inertial Navigation System with Point and Vertical Line Observations for Land Vehicle Applications. In *China Satellite Navigation Conference (CSNC) 2017 Proceedings: Volume II*; Lecture Notes in Electrical Engineering; Springer: Berlin/Heidelberg, Germany, 2017; pp. 445–457.
55. Boguspayev, N.; Akhmedov, D.; Raskaliyev, A.; Kim, A.; Sukhenko, A. A comprehensive review of GNSS/INS integration techniques for land and air vehicle applications. *Appl. Sci.* **2023**, *13*, 4819. [CrossRef]
56. El-Sheimy, N.; Youssef, A. Inertial sensors technologies for navigation applications: State of the art and future trends. *Satell. Navig.* **2020**, *1*, 1–21. [CrossRef]
57. Bird, J.; Arden, D. Indoor navigation with foot-mounted strapdown inertial navigation and magnetic sensors [emerging opportunities for localization and tracking]. *IEEE Wirel. Commun.* **2011**, *18*, 28–35. [CrossRef]
58. Vaduvescu, V.A.; Negrea, P. Inertial Measurement Unit—A Short Overview of the Evolving Trend for Miniaturization and Hardware Structures. In Proceedings of the 2021 International Conference on Applied and Theoretical Electricity (ICATE), Craiova, Romania, 27–29 May 2021.
59. Dutta, I.; Savoie, D.; Fang, B.; Venon, B.; Alzar, C.L.G.; Geiger, R.; Landragin, A. Continuous Cold-Atom Inertial Sensor with 1 nrad/sec Rotation Stability. *Phys. Rev. Lett.* **2016**, *116*, 183003. [CrossRef]
60. Guessoum, M.; Gautier, R.; Bouton, Q.; Sidorenkov, L.; Landragin, A.; Geiger, R. High Stability Two Axis Cold-Atom Gyroscope. In Proceedings of the 2022 9th IEEE International Symposium on Inertial Sensors and Systems (IEEE INERTIAL 2022), Avignon, France, 8–11 May 2022.
61. Zhang, L.; Gao, W.; Li, Q.; Li, R.B.; Yao, Z.W.; Lu, S.B. A Novel Monitoring Navigation Method for Cold Atom Interference Gyroscope. *Sensors* **2019**, *19*, 222. [CrossRef]
62. Palacios-Laloy, A.; Rutkowski, J.; Troadec, Y.; Léger, J.-M. On the Critical Impact of HF Power Drifts for Miniature Helium-Based NMR Gyroscopes. *IEEE Sens. J.* **2016**, *17*, 657–659. [CrossRef]
63. Tan, S.X.; Stoker, J.; Greenlee, S. Detection of foliage-obscured vehicle using a multiwavelength polarimetric lidar. In Proceedings of the IGARSS: 2007 IEEE International Geoscience and Remote Sensing Symposium, VOLS 1–12: Sensing and Understanding Our Planet, Barcelona, Spain, 23–28 July 2007; pp. 2503–2506.
64. Zhao, H.; Hua, D.X.; Mao, J.D.; Zhou, C.Y. Correction to near-range multiwavelength lidar optical parameter based on the measurements of particle size distribution. *Acta Phys. Sin.* **2015**, *64*, 124208. [CrossRef]
65. Xu, Z.; Yan, Z.; Li, X.; Shen, Z.; Zhou, Y.; Wu, Z.; Li, X. Review of high- precision multi-sensor integrated positioning towards intelligent driving. *Position Navig. Timing* **2023**, *10*, 1–20. [CrossRef]
66. Davison, A.J.; Reid, I.D.; Molton, N.D.; Stasse, O. MonoSLAM: Real-Time Single Camera SLAM. *IEEE Trans. Pattern Anal. Mach. Intell.* **2007**, *29*, 1052–1067. [CrossRef] [PubMed]
67. Klein, G.; Murray, D. Parallel Tracking and Mapping for Small AR Workspaces. In Proceedings of the 2007 6th IEEE and ACM International Symposium on Mixed and Augmented Reality, Nara, Japan, 13–16 November 2007; pp. 225–234.
68. Ng, Z.Y. *Indoor-Positioning for Warehouse Mobile Robots Using Computer Vision*; UTAR: Kampar, Malaysia, 2021.
69. Kim, T.-L.; Arshad, S.; Park, T.-H. Adaptive Feature Attention Module for Robust Visual–LiDAR Fusion-Based Object Detection in Adverse Weather Conditions. *Remote Sens.* **2023**, *15*, 3992. [CrossRef]
70. Gallego, G.; Delbrück, T.; Orchard, G.; Bartolozzi, C.; Taba, B.; Censi, A.; Leutenegger, S.; Davison, A.J.; Conradt, J.; Daniilidis, K.; et al. Event-Based Vision: A Survey. *IEEE Trans. Pattern Anal. Mach. Intell.* **2022**, *44*, 154–180. [CrossRef]
71. Wang, X.; Yang, D.; Wang, Z.; Kwan, H.; Chen, J.; Wu, W.; Li, H.; Liao, Y.; Liu, S. Towards realistic uav vision-language navigation: Platform, benchmark, and methodology. *arXiv* **2024**, arXiv:2410.07087.
72. Liu, S.; Zhang, H.; Qi, Y.; Wang, P.; Zhang, Y.; Wu, Q. Aerialvln: Vision-and-language navigation for uavs. In Proceedings of the IEEE/CVF International Conference on Computer Vision, Paris, France, 2–6 October 2023; pp. 15384–15394.
73. Lee, J.; Miyanishi, T.; Kurita, S.; Sakamoto, K.; Azuma, D.; Matsuo, Y.; Inoue, N. CityNav: Language-Goal Aerial Navigation Dataset with Geographic Information. *arXiv* **2024**, arXiv:2406.14240.
74. Li, T.; Zhang, H.P.; Gao, Z.Z.; Niu, X.J.; El-sheimy, N. Tight Fusion of a Monocular Camera, MEMS-IMU, and Single-Frequency Multi-GNSS RTK for Precise Navigation in GNSS-Challenged Environments. *Remote Sens.* **2019**, *11*, 24. [CrossRef]
75. Carlson, N.A. Federated Square Root Filter for Decentralized Parallel Processes. *IEEE Trans. Aerosp. Electron. Syst.* **1990**, *26*, 517–525. [CrossRef]

76. Schütz, A.; Sánchez-Morales, D.E.; Pany, T. Precise Positioning Through a Loosely-Coupled Sensor Fusion of Gnss-Rtk, INS and LiDAR for Autonomous Driving. In Proceedings of the 2020 IEEE/ION Position, Location and Navigation Symposium (PLANS), Portland, OR, USA, 20–23 April 2020; pp. 219–225.
77. Li, T.; Pei, L.; Xiang, Y.; Wu, Q.; Xia, S.; Tao, L.; Guan, X.; Yu, W. P 3-LOAM: PPP/LiDAR loosely coupled SLAM with accurate covariance estimation and robust RAIM in urban canyon environment. *IEEE Sens. J.* **2020**, *21*, 6660–6671.
78. Gao, Z.; Zhang, H.; Ge, M.; Niu, X.; Shen, W.; Wickert, J.; Schuh, H. Tightly coupled integration of multi-GNSS PPP and MEMS inertial measurement unit data. *GPS Solut.* **2016**, *21*, 377–391. [CrossRef]
79. Wang, D.; Dong, Y.; Li, Z.; Li, Q.; Wu, J. Constrained MEMS-Based GNSS/INS Tightly Coupled System With Robust Kalman Filter for Accurate Land Vehicular Navigation. *IEEE Trans. Instrum. Meas.* **2020**, *69*, 5138–5148. [CrossRef]
80. Chen, K.; Chang, G.B.; Chen, C. GINav: A MATLAB-based software for the data processing and analysis of a GNSS/INS integrated navigation system. *GPS Solut.* **2021**, *25*, 7. [CrossRef]
81. Luo, N.; Hu, Z.; Ding, Y.; Li, J.; Zhao, H.; Liu, G.; Wang, Q. DFF-VIO: A General Dynamic Feature Fused Monocular Visual-Inertial Odometry. *IEEE Trans. Circuits Syst. Video Technol.* **2024**, *35*, 1758–1773. [CrossRef]
82. Liu, C.; Jiang, C.; Wang, H. InGVIO: A Consistent Invariant Filter for Fast and High-Accuracy GNSS-Visual-Inertial Odometry. *IEEE Robot. Autom. Lett.* **2023**, *8*, 1850–1857. [CrossRef]
83. Li, X.X.; Li, S.Y.; Zhou, Y.X.; Shen, Z.H.; Wang, X.B.; Li, X.; Wen, W.S. Continuous and Precise Positioning in Urban Environments by Tightly Coupled Integration of GNSS, INS and Vision. *IEEE Robot. Autom. Lett.* **2022**, *7*, 11458–11465. [CrossRef]
84. Jiang, H.; Yan, D.; Wang, J.; Yin, J. Innovation-based Kalman filter fault detection and exclusion method against all-source faults for tightly coupled GNSS/INS/Vision integration. *GPS Solut.* **2024**, *28*, 108. [CrossRef]
85. Li, S.Y.; Li, X.X.; Wang, H.D.; Zhou, Y.X.; Shen, Z.H. Multi-GNSS PPP/INS/Vision/LiDAR tightly integrated system for precise navigation in urban environments. *Inf. Fusion* **2023**, *90*, 218–232. [CrossRef]
86. Li, X.X.; Wang, X.B.; Liao, J.C.; Li, X.; Li, S.Y.; Lyu, H.B. Semi-tightly coupled integration of multi-GNSS PPP and S-VINS for precise positioning in GNSS-challenged environments. *Satell. Navig.* **2021**, *2*, 14. [CrossRef]
87. Liu, C.L.; Wang, C.; Wang, J. A Bandwidth Adaptive Pseudo-Code Tracking Loop Design for BD/INS Integrated Navigation. In Proceedings of the 2nd International Conference on Control Science and Systems Engineering (ICCSSE), Singapore, 27–29 July 2016; pp. 46–49.
88. Liu, H.R.; Zhang, T.S.; Zhang, P.H.; Qi, F.R.; Li, Z. Accuracy Analysis of GNSS/INS Deeply-Coupled Receiver for Strong Earthquake Motion. In Proceedings of the 8th China Satellite Navigation Conference (CSNC), Shanghai, China, 23–25 May 2017; pp. 339–349.
89. Wu, M.Y.; Ding, J.C.; Zhao, L.; Kang, Y.Y.; Luo, Z.B. An adaptive deep-coupled GNSS/INS navigation system with hybrid pre-filter processing. *Meas. Sci. Technol.* **2018**, *29*, 14. [CrossRef]
90. Ruotsalainen, L.; Kirkko-Jaakkola, M.; Bhuiyan, M.; Söderholm, S.; Thombre, S.; Kuusniemi, H. Deeply coupled GNSS, INS and visual sensor integration for interference mitigation. In Proceedings of the 27th International Technical Meeting of the Satellite Division of The Institute of Navigation (ION GNSS+ 2014), Tampa, FL, USA, 8–12 September 2014; pp. 2243–2249.
91. Zuo, Z.; Yang, B.; Li, Z.; Zhang, T. A GNSS/IMU/Vision Ultra-Tightly Integrated Navigation System for Low Altitude Aircraft. *IEEE Sens. J.* **2022**, *22*, 11857–11864. [CrossRef]
92. Indelman, V.; Gurfil, P.; Rivlin, E.; Rotstein, H. Real-time vision-aided localization and navigation based on three-view geometry. *IEEE Trans. Aerosp. Electron. Syst.* **2012**, *48*, 2239–2259.
93. Bloesch, M.; Omari, S.; Hutter, M.; Siegwart, R. Robust visual inertial odometry using a direct EKF-based approach. In Proceedings of the 2015 IEEE/RSJ International Conference on Intelligent Robots and Systems (IROS), Hamburg, Germany, 28 September–2 October 2015; pp. 298–304.
94. Castellanos, J.A.; Martinez-Cantin, R.; Tardós, J.D.; Neira, J. Robocentric map joining: Improving the consistency of EKF-SLAM. *Robot. Auton. Syst.* **2007**, *55*, 21–29.
95. Mourikis, A.I.; Roumeliotis, S.I. A multi-state constraint Kalman filter for vision-aided inertial navigation. In Proceedings of the 2007 IEEE International Conference on Robotics and Automation, Rome, Italy, 10–14 April 2007; pp. 3565–3572.
96. Heo, S.; Jung, J.H.; Park, C.G. Consistent EKF-Based Visual-Inertial Navigation Using Points and Lines. *IEEE Sens. J.* **2018**, *18*, 7638–7649. [CrossRef]
97. Li, X.H.; Jiang, H.D.; Chen, X.Y.; Kong, H.; Wu, J.F. Closed-Form Error Propagation on $SE_n(3)$ Group for Invariant EKF With Applications to VINS. *IEEE Robot. Autom. Lett.* **2022**, *7*, 10705–10712. [CrossRef]
98. Weiss, S.; Scaramuzza, D.; Siegwart, R. Monocular-SLAM–based navigation for autonomous micro helicopters in GPS-denied environments. *J. Field Robot.* **2011**, *28*, 854–874. [CrossRef]
99. Novara, C.; Ruiz, F.; Milanese, M. Direct Filtering: A New Approach to Optimal Filter Design for Nonlinear Systems. *IEEE Trans. Autom. Control* **2013**, *58*, 86–99. [CrossRef]
100. Xin, S.; Wang, X.; Zhang, J.; Zhou, K.; Chen, Y. A Comparative Study of Factor Graph Optimization-Based and Extended Kalman Filter-Based PPP-B2b/INS Integrated Navigation. *Remote Sens.* **2023**, *15*, 5144. [CrossRef]

101. Li, X.; Zhang, X.; Niu, X.; Wang, J.; Pei, L.; Yu, F.; Zhang, H.; Yang, C.; Gao, Z.; Zhang, Q.; et al. Progress and Achievements of Multi-sensor Fusion Navigation in China during 2019--2023. *J. Geod. Geoinf. Sci.* **2023**, *6*, 102–114.
102. Zhu, F.; Xu, Z.; Zhang, X.; Zhang, Y.; Chen, W.; Zhang, X. On State Estimation in Multi-Sensor Fusion Navigation: Optimization and Filtering. *arXiv* **2024**, arXiv:2401.05836.
103. Kim, J.; Sukkarieh, S. 6DoF SLAM aided GNSS/INS navigation in GNSS denied and unknown environments. *J. Glob. Position. Syst.* **2005**, *4*, 120–128.
104. Chu, T.; Guo, N.; Backén, S.; Akos, D. Monocular camera/IMU/GNSS integration for ground vehicle navigation in challenging GNSS environments. *Sensors* **2012**, *12*, 3162–3185. [CrossRef] [PubMed]
105. Kalman, R.E. A New Approach to Linear Filtering and Prediction Problems Trans. *ASME. Ser. D* **1960**, *82*, 35–45. [CrossRef]
106. Lv, W.F.; Publishing, I.O.P. Kalman Filtering Algorithm for Integrated Navigation System in Unmanned Aerial Vehicle. In Proceedings of the 5th Annual International Conference on Information System and Artificial Intelligence (ISAI), Zhejiang, China, 22–23 May 2020.
107. Zhao, L.; Wang, X.; Ding, J.; Cao, W. Overview of nonlinear filter methods applied in integrated navigation system. *J. Chin. Inert. Technol.* **2009**, *17*, 46–52.
108. Zhang, W.; Sun, R. Research on performance comparison of EKF and UKF and their application. *J. Nanjing Univ. Sci. Technol.* **2015**, *39*, 614–618.
109. Shen, Z.; Yu, W.; Fang, J. Nonlinear algorithm based on UKF for lowcost SINS /GPS integrated navigation system. *Syst. Eng. Electron.* **2007**, *29*, 408–411.
110. Hu, G.; Wang, W.; Zhong, Y.; Gao, B.; Gu, C. A new direct filtering approach to INS/GNSS integration. *Aerosp. Sci. Technol.* **2018**, *77*, 755–764. [CrossRef]
111. Xu, C.; Chen, S.; Hou, Z. A hybrid information fusion method for SINS/GNSS integrated navigation system utilizing GRU-aided AKF during GNSS outages. *Meas. Sci. Technol.* **2024**, *35*, 106311. [CrossRef]
112. Zhang, T.; Yuan, M.; Wang, L.; Tang, H.; Niu, X. A Robust and Efficient IMU Array/GNSS Data Fusion Algorithm. *IEEE Sens. J.* **2024**, *24*, 26278–26289. [CrossRef]
113. Zhang, L.L.; Qu, H.; Mao, J.; Hu, X.P. A Survey of Intelligence Science and Technology Integrated Navigation Technology. *Navig. Position. Timing* **2020**, *7*, 50–63. [CrossRef]
114. Fornasier, A.; van Goor, P.; Allak, E.; Mahony, R.; Weiss, S. MSCEqF: A Multi State Constraint Equivariant Filter for Vision-Aided Inertial Navigation. *IEEE Robot. Autom. Lett.* **2024**, *9*, 731–738. [CrossRef]
115. Hesch, J.A.; Kottas, D.G.; Bowman, S.L.; Roumeliotis, S.I. Camera-IMU-based localization: Observability analysis and consistency improvement. *Int. J. Robot. Res.* **2013**, *33*, 182–201. [CrossRef]
116. Huai, Z.; Huang, G. Robocentric visual–inertial odometry. *Int. J. Robot. Res.* **2019**, *41*, 667–689. [CrossRef]
117. Eckenhoff, K.; Geneva, P.; Huang, G. MIMC-VINS: A Versatile and Resilient Multi-IMU Multi-Camera Visual-Inertial Navigation System. *IEEE Trans. Robot.* **2021**, *37*, 1360–1380. [CrossRef]
118. Lynen, S.; Achtelik, M.W.; Weiss, S.; Chli, M.; Siegwart, R. A Robust and Modular Multi-Sensor Fusion Approach Applied to MAV Navigation. In Proceedings of the IEEE/RSJ International Conference on Intelligent Robots and Systems (IROS), Tokyo, Japan, 3–8 November 2013; pp. 3923–3929.
119. Yue, Z.; Lian, B.; Tang, C.; Tong, K. A novel adaptive federated filter for GNSS/INS/VO integrated navigation system. *Meas. Sci. Technol.* **2020**, *31*, 85102. [CrossRef]
120. Xia, C.; Li, X.; Li, S.; Zhou, Y. Invariant-EKF-Based GNSS/INS/Vision Integration with High Convergence and Accuracy. *IEEE/ASME Trans. Mechatron.* **2024**, *29*, 1–12. [CrossRef]
121. Liao, J.; Li, X.; Wang, X.; Li, S.; Wang, H. Enhancing navigation performance through visual-inertial odometry in GNSS-degraded environment. *GPS Solut.* **2021**, *25*, 50. [CrossRef]
122. Gu, S.; Dai, C.; Mao, F.; Fang, W. Integration of Multi-GNSS PPP-RTK/INS/Vision with a Cascading Kalman Filter for Vehicle Navigation in Urban Areas. *Remote Sens.* **2022**, *14*, 4337. [CrossRef]
123. Li, K.; Li, J.; Wang, A. Discussion on development of GNSS/INS/Visual integrated navigation technology and data fusion. *J. Navig. Position.* **2023**, *11*, 9–15. [CrossRef]
124. Hu, J.-S.; Chen, M.-Y. A sliding-window visual-IMU odometer based on tri-focal tensor geometry. In Proceedings of the 2014 IEEE International Conference on Robotics and Automation (ICRA), Hong Kong, China, 31 May–7 June 2014; pp. 3963–3968.
125. Xian, Z.; Hu, X.; Lian, J. Fusing stereo camera and low-cost inertial measurement unit for autonomous navigation in a tightly-coupled approach. *J. Navig.* **2015**, *68*, 434–452. [CrossRef]
126. Kong, X.; Wu, W.; Zhang, L.; Wang, Y. Tightly-coupled stereo visual-inertial navigation using point and line features. *Sensors* **2015**, *15*, 12816–12833. [CrossRef]
127. Huang, G.P.; Mourikis, A.I.; Roumeliotis, S.I. A quadratic-complexity observability-constrained unscented Kalman filter for SLAM. *IEEE Trans. Robot.* **2013**, *29*, 1226–1243.

128. Xu, J.; Yang, G.; Sun, Y.; Picek, S. A Multi-Sensor Information Fusion Method Based on Factor Graph for Integrated Navigation System. *IEEE Access* **2021**, *9*, 12044–12054. [CrossRef]
129. Taghizadeh, S.; Nezhadshahbodaghi, M.; Safabakhsh, R.; Mosavi, M.R. A low-cost integrated navigation system based on factor graph nonlinear optimization for autonomous flight. *GPS Solut.* **2022**, *26*, 78.
130. Wen, W.; Pfeifer, T.; Bai, X.; Hsu, L.-T. Factor graph optimization for GNSS/INS integration: A comparison with the extended Kalman filter. *Navig. J. Inst. Navig.* **2021**, *68*, 315–331.
131. Chiu, H.-P.; Zhou, X.S.; Carlone, L.; Dellaert, F.; Samarasekera, S.; Kumar, R. Constrained optimal selection for multi-sensor robot navigation using plug-and-play factor graphs. In Proceedings of the 2014 IEEE International Conference on Robotics and Automation (ICRA), Hong Kong, China, 31 May–7 June 2014; pp. 663–670.
132. Indelman, V.; Williams, S.; Kaess, M.; Dellaert, F. Information fusion in navigation systems via factor graph based incremental smoothing. *Robot. Auton. Syst.* **2013**, *61*, 721–738.
133. Li, W.; Cui, X.W.; Lu, M.Q. A Robust Graph Optimization Realization of Tightly Coupled GNSS/INS Integrated Navigation System for Urban Vehicles. *Tsinghua Sci. Technol.* **2018**, *23*, 724–732. [CrossRef]
134. Lin, F.; Wang, S.; Chen, Y.; Zou, M.; Peng, H.; Liu, Y. Vehicle integrated navigation IMU mounting angles estimation method based on nonlinear optimization. *Meas. Sci. Technol.* **2023**, *35*, 036304. [CrossRef]
135. Zhang, L.; Wen, W.; Hsu, L.-T.; Zhang, T. An improved inertial preintegration model in factor graph optimization for high accuracy positioning of intelligent vehicles. *IEEE Trans. Intell. Veh.* **2023**, *9*, 1641–1653.
136. Li, T.; Zhang, H.; Han, B.; Xia, M.; Shi, C. Relative Accuracy of GNSS/INS Integration Based on Factor Graph Optimization. *IEEE Sens. J.* **2024**, *24*, 33182–33194. [CrossRef]
137. Walcott-Bryant, A.; Kaess, M.; Johannsson, H.; Leonard, J.J. Dynamic Pose Graph SLAM: Long-term Mapping in Low Dynamic Environments. In Proceedings of the 2012 IEEE/RSJ International Conference on Intelligent Robots and Systems (IROS), Vilamoura-Algarve, Portugal, 7–12 October 2012; pp. 1871–1878.
138. Li, J.; Pan, X.; Huang, G.; Zhang, Z.; Wang, N.; Bao, H.; Zhang, G. RD-VIO: Robust Visual-Inertial Odometry for Mobile Augmented Reality in Dynamic Environments. *IEEE Trans. Vis. Comput. Graph.* **2024**, *30*, 6941–6955. [CrossRef]
139. Zheng, F.; Lin, W.; Sun, L. Dyna VIO: Real-Time Visual-Inertial Odometry with Instance Segmentation in Dynamic Environments. In Proceedings of the 2024 4th International Conference on Computer, Control and Robotics (ICCCR), Shanghai, China, 19–21 April 2024; pp. 21–25.
140. Mascaro, R.; Teixeira, L.; Hinzmann, T.; Siegwart, R.; Chli, M. GOMSF: Graph-Optimization Based Multi-Sensor Fusion for robust UAV Pose estimation. In Proceedings of the 2018 IEEE International Conference on Robotics and Automation (ICRA), Brisbane, Australia, 21–25 May 2018; pp. 1421–1428.
141. Niu, X.; Tang, H.; Zhang, T.; Fan, J.; Liu, J. IC-GVINS: A Robust, Real-Time, INS-Centric GNSS-Visual-Inertial Navigation System. *IEEE Robot. Autom. Lett.* **2023**, *8*, 216–223. [CrossRef]
142. Chi, C.; Zhang, X.; Liu, J.; Sun, Y.; Zhang, Z.; Zhan, X. GICI-LIB: A GNSS/INS/Camera Integrated Navigation Library. *IEEE Robot. Autom. Lett.* **2023**, *8*, 7977. [CrossRef]
143. Yin, J.; Li, T.; Yin, H.; Yu, W.X.; Zou, D.P. Sky-GVINS: A sky-segmentation aided GNSS-Visual-Inertial system for robust navigation in urban canyons. *Geo-Spat. Inf. Sci.* **2023**, *11*, 2257–2267. [CrossRef]
144. Ravi, R.; Lin, Y.-J.; Elbahnasawy, M.; Shamseldin, T.; Habib, A. Simultaneous System Calibration of a Multi-LiDAR Multicamera Mobile Mapping Platform. *IEEE J. Sel. Top. Appl. Earth Obs. Remote Sens.* **2018**, *11*, 1694–1714. [CrossRef]
145. Zhou, T.; Hasheminasab, S.M.; Habib, A. Tightly-coupled camera/LiDAR integration for point cloud generation from GNSS/INS-assisted UAV mapping systems. *ISPRS J. Photogramm. Remote Sens.* **2021**, *180*, 336–356. [CrossRef]
146. Liao, J.; Li, X.; Feng, S. GVIL: Tightly—Coupled GNSS PPP/Visual/INS/LiDAR SLAM Based on Graph Optimization. *Geomat. Inf. Sci. Wuhan Univ.* **2023**, *48*, 1204–1215. [CrossRef]
147. Liu, F.; Zhao, H.; Chen, W. A Hybrid Algorithm of LSTM and Factor Graph for Improving Combined GNSS/INS Positioning Accuracy during GNSS Interruptions. *Sensors* **2024**, *24*, 5605. [CrossRef]
148. Wu, F.; Luo, H.; Zhao, F.; Wei, L.; Zhou, B. Optimizing GNSS/INS Integrated Navigation: A Deep Learning Approach for Error Compensation. *IEEE Signal Process. Lett.* **2024**, *31*, 3104–3108. [CrossRef]
149. Meng, X.; Tan, H.; Yan, P.; Zheng, Q.; Chen, G.; Jiang, J. A GNSS/INS Integrated Navigation Compensation Method Based on CNN–GRU + IRAKF Hybrid Model During GNSS Outages. *IEEE Trans. Instrum. Meas.* **2024**, *73*, 1–15. [CrossRef]
150. Liu, J.; Guo, G. Vehicle Localization During GPS Outages with Extended Kalman Filter and Deep Learning. *IEEE Trans. Instrum. Meas.* **2021**, *70*, 1–10. [CrossRef]
151. Clark, R.; Wang, S.; Wen, H.; Markham, A.; Trigoni, N. VINet: Visual-Inertial Odometry as a Sequence-to-Sequence Learning Problem. In Proceedings of the 31st AAAI Conference on Artificial Intelligence, San Francisco, CA, USA, 4–9 February 2017; pp. 3995–4001.
152. Chen, C.H.; Lu, C.X.; Wang, B.; Trigoni, N.; Markham, A. DynaNet: Neural Kalman Dynamical Model for Motion Estimation and Prediction. *IEEE Trans. Neural Netw. Learn. Syst.* **2021**, *32*, 5479–5491. [CrossRef]

153. Han, L.; Lin, Y.; Du, G.; Lian, S. DeepVIO: Self-supervised Deep Learning of Monocular Visual Inertial Odometry using 3D Geometric Constraints. In Proceedings of the 2019 IEEE/RSJ International Conference on Intelligent Robots and Systems (IROS), Macau, China, 3–8 November 2019; pp. 6906–6913.
154. Kurt, Y.B.; Akman, A.; Alatan, A.A. Causal Transformer for Fusion and Pose Estimation in Deep Visual Inertial Odometry. *arXiv* **2024**, arXiv:2409.08769.
155. Chen, C.H.; Rosa, S.; Miao, Y.S.; Lu, C.X.X.; Wu, W.; Markham, A.; Trigoni, N.; Soc, I.C. Selective Sensor Fusion for Neural Visual-Inertial Odometry. In Proceedings of the 32nd IEEE/CVF Conference on Computer Vision and Pattern Recognition (CVPR), Long Beach, CA, USA, 16–20 June 2019; pp. 10534–10543.
156. Ragab, H.; Abdelaziz, S.K.; Elhabiby, M.; Givigi, S.; Noureldin, A.; Inst, N. Machine Learning-based Visual Odometry Uncertainty Estimation for Low-cost Integrated Land Vehicle Navigation. In Proceedings of the 33rd International Technical Meeting of the Satellite-Division-of-The-Institute-of-Navigation (ION GNSS), Electr Network, 21–25 September 2020; pp. 2569–2578.
157. Ma, C.; Lin, J.; Zhao, Y.; Shi, Q.; Xia, G.; Qiu, A.; Huang, J. Low Noise Temperature Compensation Strategy for North-finding MEMS Gyroscope. In Proceedings of the 2024 IEEE SENSORS, Kobe, Japan, 20–23 October 2024; pp. 1–4.
158. Li, K.; Cui, R.; Cai, Q.; Wei, W.; Shen, C.; Tang, J.; Shi, Y.; Cao, H.; Liu, J. A Fusion Algorithm for Real-Time Temperature Compensation and Noise Suppression With a Double U-Beam Vibration Ring Gyroscope. *IEEE Sens. J.* **2024**, *24*, 7614–7624. [CrossRef]
159. Li, A.; Cui, K.; An, D.; Wang, X.; Cao, H. Multi-Frame Vibration MEMS Gyroscope Temperature Compensation Based on Combined GWO-VMD-TCN-LSTM Algorithm. *Micromachines* **2024**, *15*, 1379. [CrossRef]
160. Rodriguez Mendoza, L.; O'Keefe, K. Wearable Multi-Sensor Positioning Prototype for Rowing Technique Evaluation. *Sensors* **2024**, *24*, 5280. [CrossRef] [PubMed]
161. Yan, P.; Jiang, J.; Tan, H.; Zheng, Q.; Liu, J. High Precision Time Synchronization Strategy for Low-Cost Embedded GNSS/MEMS-IMU Integrated Navigation Module. *IEEE Trans. Intell. Transp. Syst.* **2024**, *25*, 14087–14099. [CrossRef]

MDPI AG
Grosspeteranlage 5
4052 Basel
Switzerland
Tel.: +41 61 683 77 34

Remote Sensing Editorial Office
E-mail: remotesensing@mdpi.com
www.mdpi.com/journal/remotesensing

www.ingramcontent.com/pod-product-compliance
Lightning Source LLC
Chambersburg PA
CBHW061553210926
51850CB00009B/2088
* 9 7 8 3 7 2 5 8 6 6 3 6 6 *